Decemer 1999

MONOGRAPHS ON STATISTICS AND APPLIED PROBABILITY

General Editors

D.R. Cox, V. Isham, N. Keiding, T. Louis, N. Reid, and H. Tong

1 Stochastic Population Models in Ecology and Epidemiology *M.S. Barlett* (1960)

2 Queues D.R. *Cox and W.L. Smith* (1961)

3 Monte Carlo Methods *J.M. Hammersley and D.C. Handscomb* (1964)

4 The Statistical Analysis of Series of Events *D.R. Cox and P.A.W. Lewis* (1966)

5 Population Genetics *W.J. Ewens* (1969)

6 Probability, Statistics and Time *M.S. Barlett* (1975)

7 Statistical Inference *S.D. Silvey* (1975)

8 The Analysis of Contingency Tables *B.S. Everitt* (1977)

9 Multivariate Analysis in Behavioural Research *A.E. Maxwell* (1977)

10 Stochastic Abundance Models *S. Engen* (1978)

11 Some Basic Theory for Statistical Inference *E.J.G. Pitman* (1979)

12 Point Processes *D.R. Cox and V. Isham* (1980)

13 Identification of Outliers *D.M. Hawkins* (1980)

14 Optimal Design *S.D. Silvey* (1980)

15 Finite Mixture Distributions *B.S. Everitt and D.J. Hand* (1981)

16 Classification *A.D. Gordon* (1981)

17 Distribution-free Statistical Methods, 2nd edition *J.S. Maritz* (1995)

18 Residuals and Influence in Regression *R.D. Cook and S. Weisberg* (1982)

19 Applications of Queueing Theory, 2nd edition *G.F. Newell* (1982)

20 Risk Theory, 3rd edition *R.E. Beard, T. Pentikäinen and E. Pesonen* (1984)

21 Analysis of Survival Data *D.R. Cox and D. Oakes* (1984)

22 An Introduction to Latent Variable Models *B.S. Everitt* (1984)

23 Bandit Problems *D.A. Berry and B. Fristedt* (1985)

24 Stochastic Modelling and Control *M.H.A. Davis and R. Vinter* (1985)

25 The Statistical Analysis of Composition Data *J. Aitchison* (1986)

26 Density Estimation for Statistics and Data Analysis *B.W. Silverman* (1986)

27 Regression Analysis with Applications *G.B. Wetherill* (1986)

28 Sequential Methods in Statistics, 3rd edition
G.B. Wetherill and K.D. Glazebrook (1986)

29 Tensor Methods in Statistics *P. McCullagh* (1987)

30 Transformation and Weighting in Regression
R.J. Carrol and D. Ruppert (1988)

31 Asymptotic Techniques of Use in Statistics
O.E. Bardorff-Nielsen and D.R. Cox (1989)

32 Analysis of Binary Data, 2nd edition *D.R. Cox and E.J. Snell* (1989)

Statistical Aspects of BSE and vCJD

Models for Epidemics

CHRISTL A. DONNELLY

Head of Statistics Unit
Wellcome Trust Centre
for the Epidemiology of Infectious Disease
University of Oxford, UK

NEIL M. FERGUSON

Royal Society University Research Fellow
Wellcome Trust Centre
for the Epidemiology of Infectious Disease
University of Oxford, UK.

CHAPMAN & HALL/CRC

Boca Raton   London   New York   Washington, D.C.

Library of Congress Cataloging-in-Publication Data

Statistical aspects of BSE and vCJD: models for epidemics / Christl
A. Donnelly, Neil M. Ferguson.
      p. cm. (Monographs on statistics and applied probability;
84)
    Includes bibliographical references and index.
    ISBN 0-8493-0386-9 (alk. paper)
    1. Prion diseases--Epidemiology--Statistical methods. 2. Bovine spongi-
form encephalopathy--Epidemiology--Statistical methods.
3. Creutzfeldt-Jacob disease--Epidemiology--Statistical methods.
4. Epidemiology--Statistical methods. I. Series
    RA644.P93 S73 1999
    614.5′98′015195—dc21   99-35812
                                                              CIP

© 2000 by Chapman & Hall/CRC

No claim to original U.S. Government works
International Standard Book Number 0-8493-0386-9
Library of Congress Card Number 99-35812
Printed in the United States of America   1 2 3 4 5 6 7 8 9 0
Printed on acid-free paper

Contents

CONTENTS

Preface

From the beginning of the epidemic of BSE, extensive databases have been compiled with data from every animal suspected of exhibiting the clinical signs of BSE notified to the Central Veterinary Laboratory (CVL), Weybridge England, or to the Department of Agriculture Northern Ireland (DANI). Requests from non-governmental scientists for full access to the CVL BSE database were refused until May 1996 when our research group, headed by Professor Roy Anderson, was given permission to perform independent analyses of the BSE epidemic in Great Britain.

Additional experiments and studies have been conducted to describe the transmissibility of the BSE agent and the course of the disease. A range of statistical techniques are required for the analysis of these data due to the complex nature of the disease and the absence of an ante-mortem test for infection. We derive and describe the techniques utilised to estimate parameters and to make projections of future trends in BSE and vCJD incidence both in terms of cases and infections. For each analysis discussed, we highlight the assumptions required and the limitations of available data, and put the methods used into the context of other epidemiological and statistical models of infectious disease. In addition, we touch on the political responses and implications of results where appropriate.

We warmly thank Roy Anderson, Azra Ghani, Thomas Hagenaars and David Cox without whom this book would not have been possible. We are grateful to John Wilesmith, Judi Ryan, Ann Nolan, Owen Denny, and Pedro Simas for the provision of data on BSE cases, and to Bob Will and James Ironside for the provision of data on vCJD cases. We would also like to thank Mark Woolhouse, Helen Udy, Steve Dunstan and Catherine Watt for their contributions to the research discussed in this book, and John Pattison, Peter Smith, Frank Kelly, Chris Bostock, Bob May, John Collinge, Richard Southwood, Nora Hunter, Tom Eddy, Sheila Gore,

Samantha MaWhinney and many others for helpful discussions on various aspects of this work.

We thank our families for continuing encouragement. Finally, N.M.F. would like to thank Helen, and C.A.D., Ben, for their patience and constant support in the writing of this book (and much else besides).

C.A.D. is supported as the Head of the Statistics Unit by the Wellcome Trust and N.M.F. by a Royal Society University Research Fellowship at the Wellcome Trust Centre for the Epidemiology of Infectious Disease. The Wellcome Trust and MAFF also provided grant support for much of the research that is discussed in this book.

Oxford   Christl Donnelly
June 1999   Neil Ferguson

CHAPTER 1

Introduction

1.1 Background and aims

Infectious diseases have been a major cause of mortality and thus an important force of natural selection in both animal and plant communities. With respect to humans, they remain a major public health problem in developing countries, though their impact in the developed world has been substantially reduced in the twentieth century by mass immunization and the use of anti-microbial drugs. However, continued progress is now being questioned in the face of emerging pathogens (*e.g.* HIV, new variant Creutzfeldt-Jakob Disease (vCJD), new influenza strains), and drug resistant strains of previously controlled diseases (*e.g.* tuberculosis (TB)). Similarly, novel and re-emergent zoonoses (*e.g.* bovine spongiform encephalopathy (BSE) and bovine TB) pose a significant risk to animal and human health in many societies.

Traditional descriptive epidemiological methods have been useful in identifying risk factors for both non-communicable and infectious diseases, but give limited insight into the non-linear transmission dynamics of infectious diseases in populations. On the other hand, dynamical models, traditionally used in ecology and mathematical biology, have the ability to identify the key biological and epidemiological determinants of observed pattern. Most such models share certain common features: they compartmentalize the population by disease state (into susceptible, infected and recovered individuals, for instance) and represent the epidemic process as a feedback loop whereby the rate at which susceptible individuals become infected is proportional to the number of infected individuals in the population (the so-called 'mass action principle'). It is this central density-dependent non-linearity of epidemic models that generates the complex dynamics exhibited by many disease systems. The dramatic differences in the behaviour of different host-pathogen systems lie in the precise nature of how the infection contact process is mediated by heterogeneities in the structure of the host or pathogen populations. For example, for many childhood diseases, the risk of infection is largely determined by the degree of peer-group mixing, leading to seasonal (generated by school terms) and age-dependent effects dominating the pattern of transmission – and causing the distinctive biennial epidemics seen in measles incidence time-series. It was by linking such transmission models to age-stratified morbidity data that insight was given into the potentially dangerous consequences of limited immunization programmes in increasing the incidence of congenital rubella syndrome through an increase in the average age at infection. For sexually transmitted diseases, host heterogeneity in sexual behaviour – modulated by disease biology – largely determines epidemic behaviour; models incorporating such heterogeneity were used to predict key features of the AIDS pandemic.

However, with the increasing realism and power of dynamical models have come increasing complexity and consequent difficulties in model parametrization and validation. Parameter estimation is key to any application of non-linear models, as their qualitative behaviour often changes dramatically even with small changes in parameter values. Advances in statistical techniques and computing resources have greatly extended the range of situations where the application of such methods is now feasible.

This book aims to serve as an example of how such often complex models of disease transmission dynamics can be placed into the best statistical context to allow robust parameter estimation and sensitivity analysis to be performed. The challenge is to retain the best aspects of both the biostatistics and biomathematics traditions: on one hand, the attempt to characterize interactions between key variables in as assumption-free a manner as possible, and, on the other, to construct models that describe the mechanisms underlying observed interactions. We place our discussion in the firmly applied context of the analysis of epidemiological data on the most important emergent animal and human infections to have affected the United Kingdom since HIV: the novel transmissible spongiform encephalopathies (TSEs) BSE and vCJD.

These diseases have complex pathways of transmission, long incubation periods and many biological uncertainties associated with their pathogenesis. Furthermore, the lack (until very recently) of reliable diagnostic tests for infection mean that reported cases of disease are often the sole epidemiological data available. Despite this, addressing many of the key epidemiological questions regarding the past patterns of exposure and the potential future course of the BSE and vCJD epidemics requires detailed consideration of the mechanisms of transmission. The need for mechanistic insight in the face of great uncertainty regarding key epidemiological parameters is what necessitates the melding of non-linear dynamical models with statistically robust estimation techniques — primarily survival analysis and maximum likelihood methods. Within this framework, the outcome of interest is progression to observable disease, with the failure hazard — a combination of the hazard of infection and the incubation period — being specified by some form of transmission model.

In the past, the type of analyses presented in this book would have been largely impossible to perform — not because of their mathematical difficulty, but because of the computational burden imposed by parameter estimation for complex non-linear models. Many of the results presented were obtained only after many hours of processing time on powerful multi-processor computers. Furthermore, the bulk of the work presented required the use of highly optimized, custom-written programs: modern statistical packages (*e.g. SAS, S-PLUS*), while increasingly powerful and flexible, still provided inadequate performance in analysing complex non-linear models of the type described here. That said, most readers will doubtless be pleased to learn that this book has not been written as a text on high-performance technical computing (there are many excellent books already available). However, we hope the fact that use of the type of techniques outlined here currently requires considerable programming expertise will increase the pressure on statistical software producers to improve their products' handling of non-linear models.

Despite our best efforts, there were a few instances where computational limits necessitated scaling back our ambition for the maximum possible statistical rigour: in the calculation of confidence regions around best-fit points in high-dimensional parameter spaces of complex geometry, and in the fitting of explicitly stochastic dynamical models to complex data. Both these areas are the focus of much current statistical research (*e.g.* Markov chain Monte Carlo methods), but existing techniques are still too resource intensive for application to some of the problems discussed here. For that reason, we place some emphasis on alternative methods for comparing models with data. Selective sensitivity analysis and parameter sampling (scenario analysis) are shown to be useful tools when examining the population-level epidemiology of the BSE and vCJD epidemics. We also demonstrate how more qualitative comparisons of various summary statistics − a historically popular technique in mathematical biology − can still have value when trying to gain insight into the epidemiological mechanisms underlying clustering of BSE cases in the United Kingdom.

In focussing on a specific set of applications relating to TSE epidemiology, we have deliberately structured this book as more of an exposition of how modern biostatistical methods can give insight into a set of complex datasets than as a methodological tutorial. It is therefore more intended to stimulate researchers working in related areas to consider the potential benefits of statistically robust dynamical modelling than to serve as an exact guide as to how to model infectious disease systems. While we do present the formulation of basic epidemic models in a fairly general way, many of the more complex models are quite specifically designed to capture particular (and unusual) aspects of TSE epidemiology. However, we hope that the basic approach adopted will serve as a template for others dealing with similarly complex systems, albeit with rather different heterogeneities and behaviour.

The book is also intended to stimulate cross-disciplinary interaction, and has therefore been aimed at three distinct audiences: a) statisticians who wish to know more about dynamical, mechanistic modelling, b) mathematical modellers who wish to employ more rigorous statistical methods, and c) numerate biologists and medical/veterinary scientists interested in the population dynamics and epidemiology of BSE and vCJD. As such, it is inevitable that some of the material presented will be familiar to readers from each discipline, but we hope the remainder is sufficiently novel to maintain interest.

1.2 Overview of the book

We start in Chapter 2 by briefly reviewing the essential aspects of the biology and epidemiology of BSE and vCJD and placing these diseases in the context of the broader family of TSEs. Chapter 3 outlines the sources of data available on BSE and vCJD, and on the demography of British bovine populations.

Chapter 4 develops simple models of disease transmission, illustrating how survival analysis and traditional epidemic modelling can be integrated together. The resulting models are incorporated into a back-calculation framework for the analysis of population-level BSE incidence data. The results obtained from this model structure are then presented in Chapter 5, which also concentrates on the application of sensitivity analysis to determine the robustness of results to variation in model assumptions.

We then turn from models dealing with disease epidemiology at the large (country level) scale to a statistical treatment of the underlying heterogeneities in exposure and transmission. Such heterogeneity occurs at a variety of scales, the two most important corresponding to the individual and the herd. Chapter 6 develops a class of survival models describing the heterogeneity in disease risk and survival between individuals, while controlling for herd-level variability. Such models are applied to the analysis of data relating to maternally associated enhancement of infection risk, with the aim being to distinguish the relative contributions of direct maternal transmission of the infectious agent and genetically variable susceptibility. Chapter 7 presents the results of these analyses, using data largely from two sources: the BSE maternal cohort study, and dam identification data for BSE cases.

Heterogeneity at the holding level is then considered in more detail in Chapter 8, with the emphasis shifting away from direct fitting of survival models toward a more diverse set of statistical methods designed to give insight into the strong clustering of BSE cases observed. Stochastic metapopulation simulations are introduced in Chapter 9 as a powerful tool for exploring potential mechanisms that could generate the observed heterogeneity in exposure. A key aspect of such models is that they enable the time evolution of the higher order moments in addition to the mean number of cases in subpopulations to be analysed.

Chapter 10 discusses how our analysis of BSE transmission dynamics might provide insight into the possible future pattern of the human vCJD epidemic. The emphasis is on methods that accurately characterize the impact of great uncertainty in key statistical parameters on the outcome of any attempt to project future cases, and we discuss in some detail the advantages of adopting scenario analysis methods utilizing extensive stochastic sampling of the space of unknown parameters. While the outcome of such analysis is to demonstrate how little can be said at the current time about the potential scale of the vCJD epidemic, we demonstrate the value of the techniques in examining future predictability and the design of large-scale unlinked anonymous testing programmes.

Finally, we conclude with a brief overview of areas of TSE epidemiology warranting future attention, together with a discussion of how developments in statistical methodology are likely to affect infectious disease epidemiology in the future.

CHAPTER 2

BSE and vCJD

2.1 Transmissible spongiform encephalopathies

BSE and vCJD are two examples of a group of diseases usually termed transmissible spongiform encephalopathies (TSEs), though sometimes referred to as transmissible degenerative encephalopathies (Taylor, 1991) or 'prion diseases' (Prusiner, 1982). These diseases are characterized by the formation of accumulations of protein, known as plaques, in brain tissue and neuronal loss. Research has identified some unique properties of TSEs including prolonged persistence of the infectious agent in the environment, sporadic cases of infectious disease, and the observation that inherited cases of disease are transmissible but acquired cases are not heritable (Ridley and Baker, 1996b).

Of the known TSEs, scrapie, a disease of sheep, has been most intensively studied. The disease was noted in England in the 18th century (Stamp, 1962; Brown and Bradley, 1998). Despite the long period since its recognition, relatively little is clearly understood concerning the epidemiology of scrapie and the main routes of transmission. Its endemic persistence in flocks argues for an element of horizontal transmission. Maternal transmission is also reported to occur but the evidence is limited at present.

In addition to scrapie and BSE, TSEs have been identified in a variety of mammalian species including chronic wasting disease (CWD) in deer and elk (Williams and Young, 1980, 1982, 1992; Spraker *et al.*, 1997), transmissible mink encephalopathy (TME), (Burger and Hartsough, 1965; Hartsough and Burger, 1965, Burger and Gorham, 1977, Marsh *et al.*, 1991; Marsh and Bessen, 1994) and feline spongiform encephalopathy (FSE) (Pearson *et al.*, 1991, 1992; Fraser *et al.*, 1994). Evidence linking consumption of squirrel brains with human TSE disease suggests that there may also be as yet unidentified TSE disease in squirrel populations (Berger *et al.*, 1997).

There is much evidence to suggest the existence of distinct

strains of the scrapie agent that are distinguishable by differences in incubation periods and neuropathology in sheep and mice when host genetic background is controlled for (Bruce *et al.*, 1991). Similarly, strains have been identified in the TME agent (Marsh and Bessen, 1994).

In experiments similar to those undertaken for scrapie, BSE was transmitted to mice from seven unrelated and geographically distant cattle hosts. The consistency of the pathology and incubation period observed for all seven BSE sources suggested that all seven were infected with the same TSE strain (Fraser *et al.*, 1992; Bruce *et al.*, 1994; Bruce 1996). Although experimental work of such limited scale is likely to miss rare strains, the consistency of the observed pathology of BSE in cattle throughout the current epidemic also supports the hypothesis that, as yet, strain variation is limited for BSE (Wells *et al.*, 1992). A more recent study compared samples of suspect BSE cases reported in the period 1992 to 1994 with a similar sample of confirmed BSE cases with the onset of clinical signs of disease in the period 1987 to 1989 (Simmons *et al.*, 1996). This study found no change in the distribution and severity of the vacuolation in diseased animals, again supporting the hypothesis that the BSE epidemic arose from a single stable strain of the aetiological agent. However, given the significance of strain variation in the epidemiology of scrapie, more intensive study is needed in the case of BSE, particularly in the latter stages of the epidemic where sufficient time may have elapsed to favour the evolution of different strain types.

The TSEs affecting humans include Creutzfeldt-Jakob disease (CJD), Gerstmann-Sträussler-Scheinker syndrome, kuru and fatal familial insomnia. The incidence rate of identified sporadic CJD in Great Britain has been greater since 1990, when the National CJD Surveillance Unit was established, than in previous surveillance periods extending back to 1970 (National CJD Surveillance Unit, 1998). However, it is impossible to determine whether this is due solely to improved case ascertainment or in part to a change in the underlying incidence of the disease. The median age at death for sporadic CJD cases was 65 years.

Iatrogenic CJD cases have arisen following dural mater grafts, implantation of intracerebral electrodes and the inoculation with pituitary hormone preparations due to the accidental inoculation of the agent (Brown *et al.*, 1992). Contamination during surgery may also contribute to the incidence of CJD cases identified as sporadic, as suggested by a recent study (Collins *et al.*, 1999).

In addition to sporadic and iatrogenic forms of CJD, there are also familial cases. To date, 19 different point mutations and extra basepair insertions of the prion protein gene have been associated with familial CJD (de Silva, 1996). Gerstmann-Sträussler-Scheinker syndrome and fatal familial insomnia have also been shown to be caused by mutations in the prion protein gene (Hsiao and Prusiner, 1990; Collinge *et al.*, 1989; Brown *et al.*, 1991; Medori *et al.*, 1992).

2.2 BSE and vCJD

Bovine spongiform encephalopathy (BSE) has had enormous consequences for the agricultural industry in the United Kingdom and political relationships between European member states. By 23 February 1999, there had been 174,001 confirmed cases of BSE in Great Britain, 1783 confirmed cases in Northern Ireland (NI) (28 February 1999), and 362 confirmed cases in the Republic of Ireland (28 February 1999). Other European countries have been affected, albeit to a lesser extent, with 228 confirmed cases in Portugal by 3 March 1999 and 286 confirmed cases in Switzerland by 19 February 1999.

Although the incidence of confirmed BSE cases in Great Britain peaked in 1992, public and political attention became more intensely focussed on the disease following the announcement in the House of Commons in March 1996 that the most likely explanation for 10 cases of an apparently new variant of Creutzfeldt-Jakob disease (vCJD) in humans (Will *et al.*, 1996) was exposure to the agent of BSE. At the time, it was suggested that the human infections were the result of exposure to BSE-infected tissues before regulations were introduced to prevent any part of cattle with clinical signs and specified offal from all cattle from entering human food in late 1989 (Bovine Offal (Prohibition) Regulations, 1989).

Announcement of a new human disease apparently related to exposure to BSE triggered a crisis in consumer confidence in beef and beef products throughout Europe. Beef consumption in the United Kingdom was reduced by 25% by June 1996 (compared with 1995), though consumption rates returned to their previous levels by August. Consumer reactions were even stronger elsewhere in Europe with initial drops in beef consumption of up to 50% in Germany and Italy. Cattle prices fell correspondingly with decreases of approximately 20% in the United Kingdom and 15% in the European Union in August 1996.

BSE was first diagnosed in southern England in November 1986 upon the examination at the Central Veterinary Laboratory (CVL) of the brains of two cattle (Wells *et al.*, 1987). However, it is now recognized that a clinically affected animal diagnosed with 'spongiform encephalopathy' in 1985 by Carol Richardson at CVL was suffering from BSE (BSE Inquiry, 1999). The similarity to scrapie, an endemic spongiform encephalopathy in British sheep flocks, was quickly recognized.

In the absence of immediately available information on the transmissibility of the aetiological agent of BSE between cattle, an epidemiological study of confirmed and suspected cases of BSE was initiated in June 1987 (Wilesmith *et al.*, 1988, 1991). The study findings were consistent with exposure of cattle to a scrapie-like agent in meat and bonemeal (MBM) incorporated in cattle feed as a source of protein and found no evidence for exposure to therapeutic or agricultural chemicals being the cause of BSE. A case-control study on feeding practices confirmed this initial finding (Wilesmith *et al.*, 1992a).

The first direct evidence of the transmissibility of BSE came from mice injected with brain homogenates from confirmed cases of BSE (Fraser *et al.*, 1988). Intracerebral and intravenous inoculation of calves with similar brain homogenates produced clinical signs of BSE with clinical abnormalities occurring as early as 37 weeks after inoculation (Dawson *et al.*, 1990a). Further work confirmed the transmissibility to a wide range of animals including pigs (Dawson *et al.*, 1990b), marmosets (Baker *et al.*, 1993), goats (Foster *et al.*, 1994) and sheep (Foster *et al.*, 1994) The experimental transmission of BSE by feeding (Barlow and Middleton, 1990; Fraser *et al.*, 1992; Middleton and Barlow, 1993; Foster *et al.*, 1994; Wells *et al.*, 1994; Ministry of Agriculture, Fisheries and Food, 1996b) supported the conclusions reached on the basis of epidemiological analysis that it was the recycling of BSE infection via feed containing MBM that was responsible for the epidemic in the United Kingdom.

In mid-1988 BSE was made notifiable and the isolation of BSE suspects when calving was required (BSE Order, 1988). In addition, the ban on the use of ruminant protein in the production of ruminant feed came into force on 18 July 1988. A scientific working

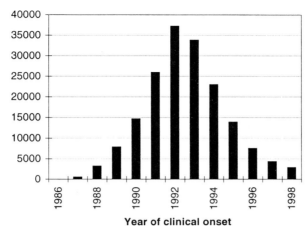

Figure 2.1 *Reported BSE cases in Great Britain by year of onset.*

party on BSE headed by Professor Sir Richard Southwood was established in May 1988 with the role of advising the Government on necessary measures in relation to both animal and human health hazards (Southwood *et al.*, 1989). On the basis of interim advice from the working party, regulations were introduced that required the destruction of carcasses of cattle affected with BSE from August 1988. Assuming that no new infections took place after the introduction of the ruminant feed ban, the working party predicted that 17,000-20,000 cases of BSE would be seen. (2160 cases had been confirmed by the end of 1988.) However, many cases of BSE arose in animals born after the introduction of these statutory measures to prevent additional feed-borne infections.

More recently, it has been estimated that feed-borne infections continued for years after the introduction of the ruminant feed ban, albeit at a much reduced rate (Anderson *et al.*, 1996). Due to the long incubation period of BSE (typically around 5 years on average (Anderson *et al.*, 1996; Ferguson *et al.*, 1997a)), the reduction in feed-borne infections did not result in an immediate reduction in the number of BSE cases arising each year. The annual incidence of BSE cases peaked in 1992 and has declined rapidly thereafter (Figure 2.1).

The BSE epidemic attracted international attention and resulted in the introduction of bans on the exportation of live British cattle to the United States and Australia in 1989 and to Japan and Canada in 1990. In addition to banning the importation of live ruminants and ruminant products from countries where BSE is known to exist, United States Department of Agriculture staff traced the cattle imported from Great Britain between 1981 and 1989 and monitor regularly all remaining imported cattle. Bans on the importation of British beef were imposed by many more countries (including Singapore, South Korea, Thailand, Philippines, New Zealand and European countries) in March 1996 following the announcement in the House of Commons.

Smaller-scale epidemics have occurred elsewhere in Europe with Northern Ireland, Switzerland, the Republic of Ireland and Portugal reporting hundreds of cases. A small number of BSE cases have arisen in other countries: France, the Netherlands, Belgium, Luxembourg, and Liechtenstein. Cases have also been reported, but only in animals imported from the United Kingdom or Switzerland, by Italy, Germany, Denmark, Oman (Carolan et al., 1990), the Falkland Islands, and and Canada.

As of 28 February 1999, 1783 cases of BSE had been confirmed in Northern Ireland, with the first BSE case having been recorded in 1988 (Denny et al., 1992; Denny and Hueston, 1997). The epidemic peaked with 56 cases per month in January 1994 and has shown a marked decline in incidence since that time (Figure 2.2).

The first BSE case in Switzerland was recorded in 1990, and by 19 February 1999, a total of 286 cases of BSE had been diagnosed (Heim and Kihm, 1999). The case epidemic peaked in 1995, five years after the introduction of a ban on the use of MBM in the production of cattle feed (Figure 2.3). It has been suggested that MBM from Great Britain is the most likely original source of the epidemic, though very little MBM was directly exported from the United Kingdom to Switzerland.

Between 1989 and 1998, 347 cases of BSE were confirmed in the Republic of Ireland (Bassett and Sheridan, 1989; Griffin et al., 1997). Only 12 of these cases were identified as imported animals. The Republic of Ireland epidemic is similar in many ways to those described above, with the majority of cases experiencing the onset of clinical signs between 4 and 6 years of age. One noteworthy feature distinguishing this epidemic is that while 14 to 19 confirmed cases were reported annually between 1989 and 1995, the incidence

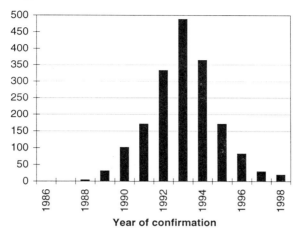

Figure 2.2 *Reported BSE cases in Northern Ireland by year of onset.*

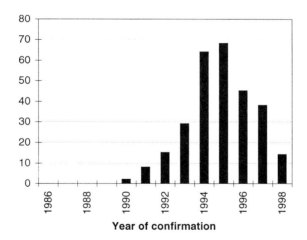

Figure 2.3 *Reported BSE cases in Switzerland by year of diagnosis.*

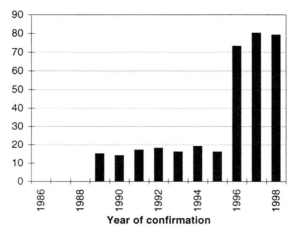

Figure 2.4 *Reported BSE cases in the Republic of Ireland by year of diagnosis.*

of clinical cases has been much greater in recent years (Figure 2.4). The reason for this increase is not clear.

An even more striking upturn in incidence has been observed in Portugal (Figure 2.5), where 228 cases have been confirmed by 3 March 1999 (7 in imported British cattle). However, analysis of the incidence data stratified by birth cohort and age at onset has detected decreasing *infection* incidence — a result of the mammalian meat and bonemeal ban introduced in June 1994 — prior to the resulting decrease in *case* incidence (Donnelly *et al.*, 1999).

In response to a recommendation by the Southwood Committee (Southwood *et al.*, 1989), national surveillance of CJD began in 1990 to identify any change in the epidemiological pattern of CJD cases that might be attributable to exposure to BSE. Since the announcement in the House of Commons in March 1996 that the aetiological agent of BSE was the most likely cause of 10 cases of an apparently new variant of Creutzfeldt-Jakob disease (vCJD) in humans (Will *et al.*, 1996), considerable additional scientific evidence in support of this hypothesis has accumulated (Will *et al.*, 1996; Collinge *et al.*, 1996; Hill *et al.*, 1997a; Bruce *et al.*, 1997). Up to the end of 1998, 39 deaths due to vCJD had been confirmed in the United Kingdom (Figure 2.6; Will *et al.*, 1999). The median age at

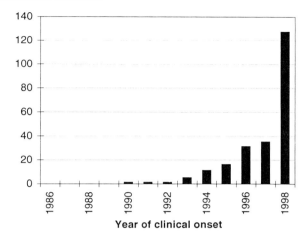

Figure 2.5 *Reported BSE cases in Portugal by year of onset.*

death was 29 years. The cases of vCJD shared a virtually indistinguishable neuropathological profile, which had not been previously described (Will *et al.*, 1996).

2.3 Transmission routes

The majority of BSE transmission is believed to have been feed-borne. The age-specific profile of feed consumption, and consequent potential exposure to the BSE agent, varies between holdings and regions, but it is likely that almost all cattle born in the United Kingdom in the 1980s were fed protein supplements that included MBM at some point during their lives (Wilesmith *et al.*, 1992a). Although the composition of feed varies according to the age of the animal, feed for animals of all ages contained MBM before the introduction of the ruminant feed ban. Calves are first given protein supplements shortly after birth, with feeding patterns being seasonal thereafter. Thus, the age-specific exposure to MBM depends on the month of birth. A minority of animals are not given supplementary feed as adults, but the feed intake of dairy cows is substantially increased at the time of first lactation and is maintained thereafter with seasonal fluctuations. The profiles of feed intake for cows on one dairy farm are shown in Figure 2.7

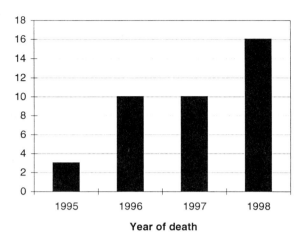

Figure 2.6 *The annual incidence of new variant Creutzfeldt-Jakob disease (vCJD) by year of death through 1998.*

for animals born in (a) December, (b) March, (c) July and (d) September. The main point emerging from these observations is that the majority of cattle were exposed to potentially contaminated feed throughout their lives and that the degree of exposure was commonly highest in adult cattle. Thus, all age classes were at risk of infection. This suggests that age-dependent changes in the susceptibility of cattle to BSE infection play a far more significant role than age-variation in feed intake in the age-dependence in infection incidence estimated in the back-calculation analysis of the transmission dynamics of BSE (Anderson *et al.*, 1996; Ferguson *et al.*, 1997a; Chapter 5).

Maternal transmission of TSEs has been a topic of research for many years. Scrapie is widely believed to be transmitted maternally and horizontally (Hoinville, 1996), though some debate continues (Ridley and Baker, 1995, 1996a). On the basis of a 4-year-old cow that experienced the onset of clinical signs of BSE in 1993 and whose dam was a confirmed BSE case, it was suggested that since this animal was born after the introduction of the ruminant feed ban, the case history was evidence of maternal transmission of the aetiological agent of BSE (Dealler and Lacey, 1994a,b). A case-control study conducted in 1994 of BSE cases born after the

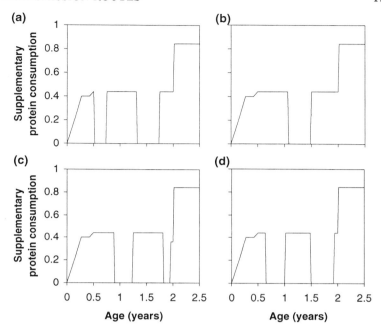

Figure 2.7 *Supplementary protein consumption for cattle on the Institute for Animal Health farm in Compton, Berkshire by month of birth ((a) December; (b) March; (c) July; (d) September).*

ruminant feed ban was consistent with maternal transmission rates between 0 and 13% (Wilesmith *et al.*, 1994; Hoinville *et al.*, 1995).

Chapter 7 summarizes analyses of the two best sources of data shedding light on the question of maternal transmission – a cohort study of the offspring of BSE-affected dams and matched controls from the same herd (Wilesmith *et al.*, 1997; Donnelly *et al.*, 1997c; Gore *et al.*, 1997; Curnow *et al.*, 1997), and dam identification data for BSE cases born after the introduction of the ruminant feed ban (Donnelly *et al.*, 1997a). The results provide clear evidence of low-level maternal transmission in the late stages of the maternal incubation period.

Maternal transmission – even at relatively high levels – cannot, on its own, ensure the long term persistence of any infection in a stable or declining population. However, in the absence of detailed knowledge of a biological mechanism or mechanisms for maternal transmission, its existence might indicate the presence of infectivity in a wider range of body tissues than has been hitherto detected (Fraser *et al.*, 1992; Wells *et al.*, 1994; Ministry of Agriculture, Fisheries and Food, 1996a). Thus, while concerns that the existence of maternal transmission may cause BSE to become endemic in British cattle in the long term are unfounded, the potentially important consequences of maternal transmission of the aetiological agent of BSE for disease biology and pathogenesis make thorough analysis of all available epidemiological data on transmission routes a priority.

Horizontal transmission of BSE by routes other than the consumption of contaminated feed has not yet been identified. Although horizontal transmission via cow-to-cow contact or a contaminated environment cannot be excluded, the time course of the epidemic following the introduction of the ban on the use of ruminant-based protein in feed suggests that if horizontal transmission does occur, it is a rare route of transmission, and that its overall contribution to transmission is on its own insufficient to maintain BSE endemically within the cattle population (see Section 5.7). However, to date no experimental studies have been published which test for the existence of horizontal transmission via contaminated pasture or cow-to-cow contact.

2.4 Incubation period distributions

TSE incubation period distributions have been shown to be strongly dependent on inoculating dose, transmission route, agent strain and host prion protein (PrP) genotype. Incubation periods have also been shown to depend on the host source of the infectious agent with serial passaging of one TSE strain within a single host species genotype tending to shorten the incubation period (Weissman, 1991a). In mice, when all these factors are held constant and intracerebral inoculations are used, the incubation periods observed tend to be extremely consistent, with the variance being small relative to the mean. When the oral transmission route is employed in mice or cattle, the variance of the incubation periods increases, but a strong dose-dependency is still seen, with highly dosed animals having much shorter incubation periods.

In a study of the survival of cattle after oral exposure to homogenized brain from cattle affected with BSE, four groups of 10 cattle were orally dosed with a specified amount of brain homogenate.

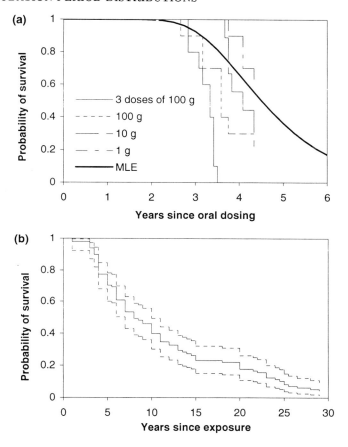

Figure 2.8 *Kaplan-Meier survival curves for (a) cattle orally dosed with 1 g, 10 g, 100 g and three doses of 100 g of brain from cattle affected with BSE (Anderson et al. 1996) with maximum likelihood estimate from the best fitting global population model (Chapter 5 and Ferguson et al. 1997a) and (b) Kaplan-Meier survival curve with 95% confidence bounds for kuru in 65 humans (Klitzman et al. 1984).*

Animals in three groups were each given a single dose of 1, 10, and 100 g with animals in the final group receiving three doses of 100 g (Anderson *et al.*, 1996). After 52 months of follow-up after dosing, BSE cases had arisen in all four dose groups (Figure 2.8(a)) with the incubation periods differing significantly between the groups.

These BSE data reveal a relatively long time delay following infection during which no cases are seen, a feature seen in many TSE incubation period distributions. No such time delay was observed in the data available on incubation periods of kuru in humans (Figure 2.8(b); Klitzman *et al.*, 1984), perhaps because the recorded incubation periods represent minimum bounds. Kuru was identified in the 1950s in a number of adjacent valleys in the mountainous interior of Papua New Guinea (Gajdusek and Zigas, 1957; Gajdusek 1977). The disease was transmitted through cannibalistic funeral rites and anthropological evidence suggests that some individuals may have received very large infectious doses, a possible explanation for the absence of a time delay in the incubation periods observed.

2.5 The genetics of TSEs

It is well known that host genotype has a major influence on susceptibility to TSEs and the typical duration of the incubation period of the disease following infection by a specific route (Spongiform Encephalopathy Advisory Committee, 1995; Prusiner *et al.*, 1996). For example, the incubation period in mice injected with BSE-infected material is strongly dependent on the PrP genotype of the host (Fraser *et al.*, 1992).

Genetic studies of scrapie infection have identified the PrP gene (which used to be called *Sip* and *Sinc* in sheep and mice, respectively, see Moore *et al.* (1998)) as having a major influence on the duration of the incubation period (Dickinson *et al.*, 1968; Hunter *et al.*, 1989, 1992; Goldmann *et al.*, 1991; Bruce *et al.*, 1994; Farquhar *et al.*, 1994). Experimental studies of scrapie have also indicated that some host genotypes appear to be resistant to infection (Davies and Kimberlin, 1985; Foster and Dickinson 1988). In a closed flock of sheep in which scrapie is endemic, Hunter and colleagues (1996) have also documented disease-linked PrP gene polymorphisms and associations of such polymorphisms (at codons 136, 154 and 171 of the PrP gene) with susceptibility to infection and the typical duration of the incubation period.

The evidence of an important role for host genotype in BSE infection and pathogenesis in cattle is less clear cut. Hunter *et al.* (1994) genotyped 370 animals for the PrP gene, 172 of which subsequently developed BSE, and failed to find an association between genotype and BSE. However, using a different technique, Neibergs *et al.* (1994) found that BSE-affected cattle and their relatives are more likely to have a particular homozygous genotype than unrelated non-BSE animals of the same breed or animals of different breeds. Epidemiological analysis of the fate of offspring of affected dams and sires may augment these limited data to shed light on the question of a genetically variable susceptibility to BSE.

Some progeny analyses have been undertaken (Wijeratne and Curnow, 1990; Curnow *et al.*, 1994; Curnow and Hau, 1996; Hau and Curnow, 1996) to examine the possibility of two classes of animals (susceptible and resistant) within five Holstein-friesian pedigree herds. However, the power of these studies was not sufficient, given the relatively small numbers of cases involved, to distinguish between alternative genetic models. The models adopted also tended to be based on genetic models with extreme assumptions such as complete resistance in a large fraction of the British cattle herd. Molecular genotypic analysis of biopsy material collected from the animals in the maternal cohort study should shed further light on the issue. However, until more detailed data are available, the ability of statistical analysis to shed greater light on the bovine genetics of BSE may be restricted to estimating an upper bound on the magnitude of genetic heterogeneity in BSE susceptibility within the British cattle population.

2.6 The nature of the aetiological agent

TSEs, including BSE and vCJD, are caused by unconventional infectious agents and manifest as neurodegenerative disease in mammalian hosts. Research into scrapie excluded bacteria, mycoplasmas and viroids as the infectious agent. The infectious agents are resistant to chemical and physical procedures that inactivate conventional microorganisms (Taylor, 1992, 1994, 1996) as well as to ultraviolet irradiation and freezing (Stamp, 1967). Formalin has been shown to be ineffective at inactivating scrapie in sheep and rodents (Taylor, 1992; Gordon *et al.*, 1940; Brown *et al.*, 1990), as well as BSE (Fraser *et al.*, 1992), CJD (Tateishi *et al.*, 1980) and TME (Burger and Gorham, 1977). Scrapie-infected brain retained infectiousness after burial for three years (Brown and Gajdusek, 1991) and scrapie infectivity is believed to remain on pasture for several years (Palsson, 1979). Similarly, CJD infectivity remained after 28 months at room temperature (Tateishi *et al.*, 1987).

There are three main interpretations of the evidence to date on the infectious agents. The infectious agents have been hypothesized to be *prions*, *virinos*, or viruses. A unified theory has also been proposed linking the virino and prion hypotheses (Weissmann, 1991b).

The term *prion* was introduced for the infectious agent of scrapie, CJD, kuru and Gerstmann-Sträussler-Scheinker syndrome to distinguish it from viruses and viroids (Prusiner, 1982). These neurodegenerative diseases are characterized by modification of the prion protein. To date, no nucleic acid has been found in prions and it is suggested that the transmissible prion is composed primarily, if not entirely, of the abnormal protein PrP^{Sc} (Alper *et al.*, 1966, 1967; Riesner *et al.*, 1992; Prusiner, 1996). The transition to the abnormal isoform (PrP^{Sc}) is a post-translational event and its presence in the host reflects the presence of infection (the prion hypothesis) (Prusiner *et al.*, 1996). Recent work has found conditions in which recombinant human PrP could switch between its native conformation and a conformation characteristic of PrP^{Sc} providing a molecular mechanism for prion propagation (Jackson *et al.*, 1999).

The prion theory can explain why no immune response has been detected in animals affected with TSEs, since immune responses would not be triggered by such a protein. The theory is supported by the finding that infectivity titre is associated with levels of the abnormal protein PrP^{Sc}. However, the number of PrP^{Sc} molecules has been observed to be orders of magnitude greater than the infectious units (Prusiner *et al.*, 1982), and Kimberlin (1990) observed that the replication of infectivity precedes the accumulation of detectable PrP^{Sc}.

One of the main concerns about the prion theory has been that it did not explain the existence of strains. However, it has recently been proposed that the self-propagation of PrP^{Sc} with distinct three-dimensional structures results in the observed strain structure of scrapie (Caughey and Chesebro, 1997). Alternatively, non-PrP chaperones may be involved in the formation of strains.

The existence of strains has been regarded by some as evidence of an informational, replicating molecule, presumably composed of nucleic acid. The presence of nucleic acid would not necessarily require a virus. It has been suggested that the infectious agent could be a core of nontranslated nucleic acid associated with cellular proteins, known as a virino (Kimberlin, 1982; Dickinson and Outram, 1998; Hope, 1994). It is further suggested that the host protein covering the nucleic acid may make the agent 'sticky' thus explaining the failure to date to purify the agent (Kimberlin, 1982).

The main reason that the slow-virus explanation was doubted was the observed resistance to inactivation of infectivity. It has been argued, however, that this resistance is limited to small subpopulations, and conventional viruses provide examples of similar resistant subpopulations (Rohwer, 1984a,b). Under the virus hypothesis, familial disease could arise due to genetically variable susceptibility to viruses.

A range of other hypotheses have been put forward as the cause of TSEs, though none has any significant supporting scientific evidence. For example, Purdey (1992, 1994, 1996a,b) suggested that the cause of BSE was chronic exposure to high doses of organophosphate pesticides, but his analyses were not sufficiently rigorous to eliminate feed containing MBM as the infection route. Other investigators have to date found no evidence for exposure to therapeutic or agricultural chemicals being the cause of BSE. Clearly, the widespread acceptance of the prion hypothesis will require the demonstration that synthesized PrP^{Sc} can be transmitted to cause disease.

2.7 Conclusion

This chapter has provided a brief overview of some of the key features of TSEs. In the following chapters, we shall see how the complex biology of this class of diseases poses particular challenges to quantitative studies of their epidemiology. In particular, their multiple transmission routes, long incubation periods and strain−structure/host genetic interactions often necessitates the use of relatively complex infection and transmission models. On the other hand, the lack (until very recently) of any test for preclinical infection means that such models tend to only be testable against limited data − from reported cases of clinical disease and a few key epidemiological experiments. Squaring the circle of being able to draw robust statistical conclusions regarding the epidemiology of TSEs on the basis of limited data is the topic of the rest of this book.

Sources of data

3.1 Introduction

In this chapter we describe the sources of the data to be analysed in following chapters, namely BSE incidence data from a variety of European countries, demographic data on British cattle, data from the maternal cohort study, and data on vCJD cases. Other data used to inform the choice of model assumptions and distributional forms are not discussed here, having already been reviewed in the previous chapter.

3.2 BSE case databases

3.2.1 The Great Britain case database

The progression of the BSE epidemic in cattle in Great Britain has been described by CVL staff (Wilesmith *et al.*, 1991, 1992a,b) via surveillance and the construction of an extensive epidemiological database recording the demographic details of diseased animals. From the beginning of the epidemic, all BSE suspects reported in Great Britain have been entered into this database, although prior to 21 June 1988 the reporting was not compulsory (BSE Order, 1988). Post-mortem diagnosis of BSE has always been based on the examination of the brain pathology. The variables recorded in the case database include herd, holding and county of diagnosis, holding and county of origin, date of birth, date of onset of the clinical signs of BSE, pregnancy status and adult herd size for herd of diagnosis, on the basis of a questionnaire completed by a veterinary officer for every BSE suspect. Annual birth cohorts were defined so that, for example, the 1989 cohort consists of cattle born between 1 July 1988 and 30 June 1989.

As with any large database, some variables are missing for some cases. Where the date of birth was missing, its value was estimated from the reported age and date at onset if these values were available. Similarly, missing dates of onset were estimated from the date of birth and age at onset. If date and age of clinical onset were both missing, the date of clinical onset was estimated to be one month prior to the completion of the BSE preliminary report by the veterinary officer.

When dates were recorded, but with only the month and year, the dates were entered into the database as the first day of the month. Figure 3.1(a) shows the loading of dates on the first day of the month for reported dates of birth, onset and purchase due in part to these approximations. No other trends are apparent in the reported dates of birth and purchase, suggesting that the person reporting either knew the date precisely or estimated it as the first of the month. In light of the difficulty in precisely identifying the day on which clinical signs of disease onset, it is not surprising that more date estimation is evident in the reported date of onset with the increased proportion of dates reported as days 7, 10, 15, 20, 25 and 30.

A much more important source of data bias arises from BSE cases born prior to July 1988 that did not have known dates of birth. For such animals, farmers provided estimated ages at onset in months, but these estimates were biased toward integer numbers of years (Figure 3.1(b)). This distorts the age-at-onset distribution of cases in early cohorts, and contributes to the observed excess of cases in each first half year of age-at-onset. These biases make the utilization of finer than yearly stratification of the incidence data problematic (Ferguson et al., 1997a). Even when yearly stratification is used, some biases remain, as it is likely that farmers might have tended to round down animals' ages to the nearest year. Rounding down ages of animals with unknown dates of birth can result in underestimation of infection numbers in the early years of the epidemic due to animals being mistakenly attributed to more recent birth cohorts. To reduce this source of bias, a stochastic resampling procedure was utilized to randomize the month of the reported age-at-onset for cases with onset dates prior to 1991 and for which no dates of birth were given (Ferguson et al., 1997a).

The analysis of finely stratified incidence data is also complicated by the significant seasonality in month of onset. There is a distinct deficit of cases following Christmas, and an excess in September-October. This trend may well be related to the degree to which cattle are under close observation by farmers, but it is non-trivial to correct.

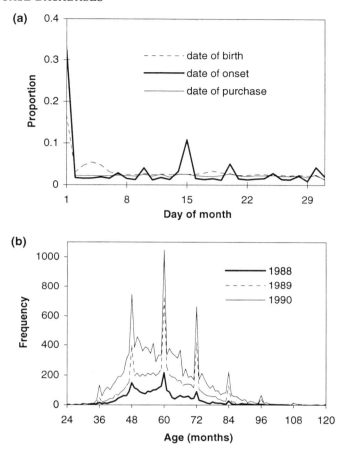

Figure 3.1 *(a) Proportion of recorded dates of birth, onset and purchase by day of the month and (b) the distribution of recorded ages in months by year of onset among confirmed BSE cases in Great Britain.*

The lowest level of aggregation of recorded BSE cases in the main database is the herd. The mean size of a herd with at least one BSE case is 81 adult animals. Herds are organized into holdings, a unit that most often corresponds to the farm. These are the basic units of analysis used in census data and are often classified into dairy and beef holdings (Section 3.3).

The age at onset for confirmed BSE cases was known for 97% of the total reported BSE cases in the main database. The distribution of age at onset differs by birth cohort as shown in Figure 3.2. It is well known that if the force of infection increases over time, the average age at infection decreases, and thus the average age at onset of clinical signs decreases (Anderson and May, 1991, page 72). Such patterns could thus have arisen even in the absence of a time-varying incubation period, given that the volume of infected material entering cattle feed was rising up through 1989. However, such a mechanism for the changing age-at-onset distribution relies on the heterogeneity in susceptibility with age being minor, an assumption that is unlikely to be valid for BSE. This is further discussed in Chapter 5 and Ferguson et al. (1997a).

Information on the BSE status of dams, sires, twins and offspring is available for some of the confirmed BSE cases in cattle born following the introduction of the ruminant feed ban in July 1988. For each of these cases, a record was made of whether the relative had been identified. For each relative identified, we were able to determine its BSE status using the main database of confirmed BSE cases. To date, 1752 dam−calf pairs of confirmed BSE cases have been identified along with 27 sire−calf pairs and 35 sets of twins.

3.2.2 The Northern Ireland case database

From June 1988, all cattle suspected of exhibiting the clinical signs of BSE in Northern Ireland (NI) either on the farm or in the abattoir were required to be reported to the Department of Agriculture for Northern Ireland (DANI). A database is maintained by DANI staff containing information on the date of birth, date of disease onset, natal herd, natal herd size, onset herd and country of origin (if not NI born) for all confirmed BSE cases in NI. The natal herd data were much more complete in NI, with natal herd data missing for only 18 confirmed cases that were not known to be imported, compared with Great Britain, where some 29% of cases are missing such data. The majority of confirmed BSE cases reported in NI are from animals born in NI − 84 cases had natal herds in Great Britain and 5 cases had natal herds in the Republic of Ireland.

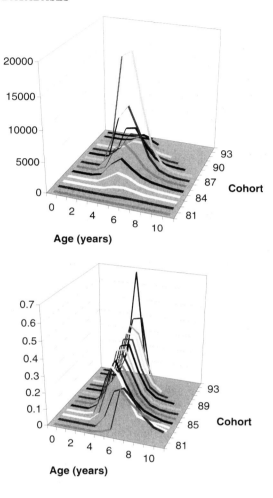

Figure 3.2 *The distribution of reported BSE cases in Great Britain by birth cohort and age at onset. (a) The incidence of cases by age at onset (in years) and birth cohort, (b) The proportion of cases arising at each age at onset (in years) by birth cohort.*

3.2.3 The Portugal case database

A database is maintained by the Portuguese National Laboratory
of Veterinary Research on all cases of BSE confirmed in Portugal.
Data are compiled on each case (dates of birth, clinical onset, cull
and confirmation; breed; geographic region) and made available on
the internet at http://www.min-agricultura.pt/.

3.3 Demography of British cattle

Given the long incubation period of BSE (around 5 years on average) and the small fraction of animals surviving beyond 3 years
(Anderson *et al.*, 1996), it is clear that the number of cattle infected
with BSE is much greater than the number of confirmed cases of
the disease. Hence, obtaining robust estimates of past trends in
the incidence of infection (given knowledge of cases of disease over
time and the incubation period distribution) requires a detailed
knowledge of the demography of the cattle population.

Annual agricultural census data with limited information about
the age distribution of cattle and beef/dairy classification were
made available to our research group on an individual holding basis. The census records, at 30 June each year, the number of cattle
in the age classes 0-6, 6-12, 12-24 and over 24 months, and summary statistics are published annually (Ministry of Agriculture,
Fisheries and Food, 1975-1990, 1992-1995; Department of Agriculture and Fisheries for Scotland, 1975-1980, 1981-1990; Scottish
Office, 1991-1995). Over the period 1974 to 1996, the national herd
has declined from 13.6 to 10.2 million cattle while the number of
calves under 12 months old (a measure of annual birth rate) has
declined from 4.0 to 2.8 million (See Figure 1(c) in Anderson *et al.*
(1996)).

Although the annual census does not provide information on the
ages of cattle over 2 years old, more detailed age profiles were obtained from the National Milk Records for a subset of dairy herds
in each of the years 1982, 1988, 1989, 1991 and 1994 with a minimum sample size of 2000 cattle. These records ignore males (which
make up a negligible fraction of the population over 2 years of
age) and do not include beef herds (the age distribution of breed-
ing beef cattle gives average ages typically up to 6 months older
than for dairy cattle). The majority of BSE cases occur in female

cattle (99.7%) and in dairy herds (80.2%, with an additional 6.8% in mixed herds).

Cattle in the National Milk Records surveys are aged by number of lactations. Data on calving intervals (Esslemont, 1992) give the average age of first lactation at 26 months with an average lactation interval of 12.5 months. This allowed the age structure to be rescaled to 1-year intervals by linear interpolation and numbers in each lactation class in each year to be weighted to give total number of cattle over 2 years recorded in the annual census for that year.

This procedure generated five age distributions. These were used for standard life table analysis (Southwood, 1978; Donnelly *et al.*, 1997b). Geometric means were used to interpolate between the estimated survival probabilities (Figure 3.3). As sufficient data were available only for cattle up to $12-13$ years old, annual mortality rates for older cattle (up to $17-18$ years) were obtained by extrapolation of the linear trend from age classes $3-4$ to $12-13$ years. Some mortality of calves up to 1 year old is ignored by this analysis. In practice, there is significant mortality of young calves (natural deaths, disposal of unwanted animals, slaughter or export as veal), but these animals are of limited relevance to interpretation of the BSE epidemic pattern.

The resulting demographic model has a variable birth rate, estimated to be the numbers of $0-1$-year-old calves obtained from the annual census data, and a survivorship function that depends upon age but not time of birth. The survivorship function shows a substantial loss of cattle between $1-2$ and $2-3$ years old (Figure 3.3), corresponding to the age at which many animals are slaughtered for beef. The mean life expectancy at birth is 3.0 years, but the mean life expectancy of cattle that are to be kept for milk production is 5.8 years. The estimated culling rates of cattle over 2 years old, $22-23\%$ per year, are in good agreement with culling rates observed in other studies (Esslemont, 1992). Over the period of the BSE epidemic, the demographic model predicts the total cattle population over 2 years old in Britain to within 4% of observed values and predicts the mean age of this population to within 0.2 years.

In contrast to the decline in the cattle herd in Great Britain, the size of the Northern Irish cattle herd has changed little over the period 1974 to 1996. While the size of the NI herd is approximately 15% of the size of the herd in Great Britain in 1995, the total

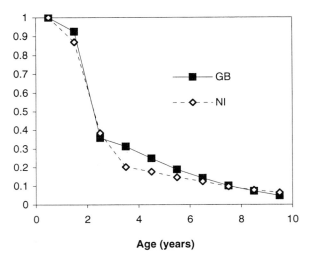

Figure 3.3 *Estimated survival probabilities of cattle in Great Britain and Northern Ireland.*

number of BSE cases in NI is only roughly 1% of the number in Great Britain. Data were available on the age structure of the NI herd at April 1996 and December 1997 by time of birth (in 4-month intervals). The survival distribution estimated from the National Milk Records data in Great Britain fit these data poorly (Ferguson *et al.*, 1998) so the survival distribution for NI was estimated using maximum likelihood methods fitting to the data on age structure of the NI herd (Figure 3.3). Survival probabilities beyond 9.5 years could not be estimated independently for NI so the conditional survival probabilities estimated for Great Britain were utilized.

Monthly calving data were also available from the National Milk Records for a subset of dairy herds in Great Britain from 1985 to 1995. The data reveal that calving was extremely seasonal in the late 1980s but became less so in the 1990s (Donnelly *et al.*, 1997b). Calving seasonality estimated from the pregnancy data on confirmed BSE cases revealed even less seasonality than that observed in the National Milk Records data over the same birth years.

Table 3.1 *Contingency table of the pair outcomes from the maternal cohort study in terms of histopathological examination of brain tissue and the onset of clinical signs of disease.*

Control	Maternally exposed			
	Hist. + Signs	Hist. + No signs	Hist. − No signs	Total
Hist. +/Signs	6	0	6	12
Hist. +/No signs	0	0	1	1
Hist. −/No signs	27	9	252	288
Total	33	9	259	301

3.4 The maternal cohort study

A cohort study of maternally associated risk factors for BSE was initiated in July 1989 (Wilesmith *et al.*, 1997). It was designed as a matched exposed-control study to determine whether the risk of BSE was higher in the offspring of BSE-affected dams after controlling for exposure to potentially infectious feed. Maternally exposed animals were recruited from the offspring of dams identified in the main BSE database for Great Britain. For each, a control animal was recruited that was born in the same herd during the same calving period to a cow that reached at least 6 years of age without developing clinical signs of BSE (but was not necessarily 6 years of age when calving). Both animals in a pair were recruited into the study at the same time. Of the 315 pairs originally recruited into the study, 301 remained after the exclusion of pairs if either animal died before March 1990 or if the dam of the control animal developed BSE at any time. Animals in both arms of the study were exposed to potentially infectious feed prior to recruitment into the study and on at least one of the three farms used for study animals after recruitment.

Table 3.1 shows the basic results of the study in terms of the disease status outcomes of the animal pairs. It should be noted that 10 of the calves diagnosed as positive were asymptomatic until slaughter at approximately 7 years of age, but histopathological examination revealed brain lesions consistent with BSE disease.

Considering only the histopathology results, the data illustrate a significantly enhanced risk of BSE among the offspring of affected dams compared with their matched controls ($p < 0.001$). The significant maternally enhanced risk of BSE may have arisen solely due to direct maternal transmission of the aetiological agent of BSE, with an estimated probability of maternal transmission, β_M, of 0.096 (95% CI: 0.051,0.142). An alternative explanation is genetically variable susceptibility to BSE infection, in which case the estimated relative risk, 5.14 (95% CI: 2.29,11.56), is an indicator of the level of heterogeneity of susceptibility (although a detailed genetic model is needed to relate the relative risk to the relative susceptibilities of the underlying genotypes (Ferguson *et al.*, 1997b)). The epidemic models required to distinguish between these two effects are developed and applied in Chapters 6 and 7.

3.5 Data on confirmed vCJD cases

The Department of Health issues monthly reports of the numbers of deaths of definite and probable cases of CJD, including sporadic, iatrogenic, familial and new variant CJD as well as Gerstmann-Sträussler-Scheinker syndrome. The number of referrals to the National CJD Surveillance Unit is also reported on a monthly basis. However, such data must be interpreted cautiously since roughly half of past referrals were found not to be CJD cases. In addition, the National CJD Surveillance Unit produces annual reports on CJD in the United Kingdom with details on the dates of clinical onset, notification to the Surveillance Unit, death and confirmation for each vCJD case (*e.g.* National CJD Surveillance Unit, 1998). The geographic distribution of the vCJD cases by place of residence at onset is also given, though no evidence of spatio-temporal clustering of cases has been found (Cousens *et al.*, 1999).

In Chapter 10 we analyse the temporal pattern and age structure of the 39 deaths due to vCJD that occurred up to the end of 1998 (Figure 3.4). Temporal stratification finer than one year was not adopted as it reveals a distinct seasonal trend in incidence (the causes of which are not yet understood).

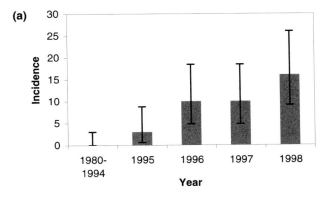

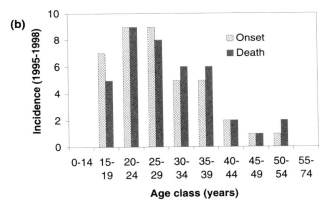

Figure 3.4 *(a) Observed annual deaths due to vCJD and exact 95% confidence intervals for the underlying incidence (assuming the incidence counts are Poisson distributed); (b) the ages at onset and death of the 39 deaths due to vCJD that occurred up to the end of 1998 (Robert Will, personal communication).*

CHAPTER 4

Population models: formulation

4.1 Introduction

A key barrier to a detailed understanding of the population biology of BSE is the lack of information on the time of infection and incubation period of individual animals. *A priori*, this makes interpretation of trends in incidence data with age and time (Figure 3.2) problematic: the epidemic curves for individual birth cohorts are superficially consistent with both the scenario of infection risk peaking at 5−6 years of life and a short incubation period, or infection risk peaking soon after birth and a long and variable incubation period. Due to the rapidly declining survivorship of older cattle, these two scenarios would lead to widely differing estimates of, for instance, the total number of cattle infected with the BSE agent throughout the epidemic. While our knowledge of other TSEs, together with limited data from the variety of epidemiological studies, would tend to favour the latter scenario, this illustrates a more general point: if little is known about the biology or epidemiology of a pathogen, the consequent uncertainty in estimates of key epidemiological parameters (*e.g.* incubation period, age-dependent susceptibility/exposure) necessitates the use of a statistically rigorous framework to assess the extent to which any particular model of transmission dynamics is consistent with the incidence data. This chapter develops such a framework, based on the back-calculation methodology developed in the early years of the HIV epidemic (Brookmeyer and Gail, 1986, 1988; Isham, 1989; Bacchetti *et al.*, 1993).

The basic approach of back-calculation models is simple: given knowledge of the incubation period distribution (IPD) and survivorship of infected individuals, the past pattern of *infection* incidence is reconstructed by deconvoluting the past pattern of *disease* incidence with the IPD, correcting for survivorship. Typically, however, knowledge of the IPD is incomplete, meaning that a variety of IPDs need to be assessed against the incidence data. In the case of

BSE, we also test a variety of other model assumptions, including the form of any age-dependent susceptibility or exposure to BSE, and the rates of maternal and direct horizontal transmission.

The models developed in this chapter are 'mean-field' approximations (Durrett and Level, 1994a,b); *i.e.* they take no account of spatial or temporal covariances, and assume that the infection hazard experienced by all animals in the population is identical. In terms of a model that explicitly represents the recycling of infection implicit in any epidemic process, this implies that all animals are 'mixing' homogeneously — the mass-action hypothesis (Hamer, 1906; Kermack and McKendrick, 1927; Anderson and May, 1991, page 7; De Jong *et al.*, 1995) — or, equivalently, every animal experiences the infection hazard generated by the mean infectivity calculated across all animals in the population. We will see in later chapters that while this is a crude approximation at best — due to the significant clustering of cases seen at the holding and county level (Chapter 8) — the population-level estimates produced by such models are remarkably similar to those produced by more sophisticated approaches.

We start the discussion by reviewing simple infection models — expressed in the mathematical framework of survival analysis — then outline how these can be extended to take account of a range of transmission routes, and a variety of susceptibility classes (representing genetically variable susceptibility to BSE). The resulting somewhat complex model is computationally costly to solve in full, so we describe techniques to simplify model evaluation, particularly in calculating the contribution of low-level maternal and horizontal transmission, and avoiding having to model explicitly the recycling of infectivity in feed while estimating the past pattern of infection. The incubation period is then introduced as a way of linking the derived mechanistic models of infection hazard to the measured outcome variable in the incidence data — namely date of clinical diagnosis. The problems inherent in calculating goodness-of-fit statistics and confidence limits for such non-linear back-calculation models are discussed.

4.2 The simple epidemic process

Consider first an epidemic occurring in a closed population of N individuals, all of whom were born at time t_0. Let $p_I(a|t_0)$ be the cumulative probability that an individual is infected by age a

(conditional on survival to that age). If $Q(t)$ is the infection hazard (or *force of infection* (Anderson and May, 1991, page 63)) at time t — defined to be the per-capita, per-unit-time probability of infection for a susceptible individual — then

$$\frac{\mathrm{d}p_I(a|t_0)}{\mathrm{d}a} = [1 - p_I(a|t_0)]\, Q(t_0 + a). \qquad (4.1)$$

Hence

$$
\begin{aligned}
p_I(A|t_0) &= \int_0^A Q(t_0 + a) \exp\left[-\int_0^a Q(t_0 + a')da'\right] da \quad (4.2) \\
&= 1 - \exp\left[-\int_0^A Q(t_0 + a')da'\right] \\
&\simeq \int_0^A Q(t_0 + a')da' \qquad (4.3)
\end{aligned}
$$

where (4.3) holds for $p_I(A|t_0) << 1$, *i.e.* when infection is relatively rare.

For future clarity, let us define

$$\rho(a|t_0) = \frac{\partial p_I(a|t_0)}{\partial a},$$

the probability density function (PDF) for infection at age a given birth at time t_0 (conditional upon survival). Note that $\rho(a|t_0)$ is also the age-specific per-capita incidence for individuals born at time t_0.

Up to this point, we have merely outlined a simple survival model. To turn this into a (non-linear) *transmission* model we make the key step of deriving an expression for the infection hazard, $Q(t)$, which captures the mechanistic process of transmission of infection between an infected and susceptible host. The simple case is for a directly (horizontally) transmitted infection. In the absence of mortality, the infection hazard at time t is just proportional to a weighted sum over all infected hosts in the population at that time, where the weighting function just gives the expected infectivity of each host at that time:

$$Q(t) = \int_0^{t-t_0} \beta_H \Psi_H(\tau)\rho(t - t_0 - \tau|t_0)d\tau \qquad (4.4)$$

where $\Psi_H(\tau)$ represents the expected standardized infectivity of an individual at time τ since infection, and β_H is the transmission coefficient for the direct horizontal route. In deriving this expression we have used the fact that $\rho(t - t_0 - \tau|t_0)$ is the density of hosts born at time t_0 who were infected time τ ago.

For simplicity, $\Psi_H(\tau)$ is standardized to have a maximum value of 1, and β_H is thus the probability per unit time of a susceptible individual becoming infected when mixing with a population of individuals all with unit infectivity. The equation (4.4) merely represents the idea that the infection hazard at time t is determined by the mean infectivity of all individuals in the population at that time. It assumes 'true' mass-action (De Jong *et al.*, 1995): that for constant population density, the number of contacts an individual has with other individuals is independent of total population size, so the probability of infection per unit time is merely the product of the number of contacts per unit time, the transmission probability per contact, and the fraction of the population infectious.

Note that in the special case that $\Psi_H(\tau) = \exp(-\eta\tau)$, an exponentially distributed infectious period, and writing $x(a) = 1 - p_I(a|t_0)$ and $y(a) = Q(t_0 + a)/\beta$ for, respectively, the fraction of the population susceptible and infectious at age a, we obtain the following equations:

$$\frac{dx}{da} = -\beta_H xy,$$
$$\frac{dy}{da} = \beta_H xy - \eta y. \qquad (4.5)$$

This is known as the SIR (susceptible-infected-removed) or Kermack–McKendrick epidemic model (Kermack and McKendrick, 1927) and underlies the great majority of the more sophisticated epidemic models used today (Anderson and May, 1991). It is worth noting that an epidemic process can be modelled as a set of first-order ordinary differential equations only if it is assumed to be a Markov process; *i.e.* the system is memoryless. While this approximation is reasonable for many short incubation period diseases, such as measles (though see Keeling and Grenfell (1997)), it is clearly inadequate for long (and possibly dose-dependent) incubation period diseases such as BSE.

The system described by (4.2) and (4.4) will experience epidemic growth only if the basic reproduction number, $R_0^{(H)}$ (the number of secondary infections generated by one primary 'typical' infection in an entirely susceptible population) is greater than unity. Here,

$R_0^{(H)}$ is given by

$$R_0^{(H)} = \beta_H \int_0^\infty \Psi_H(\tau)d\tau$$

since $\beta_H \Psi_H(\tau)/N$ is the per-capita, per-unit-time infection probability for susceptible individuals in a population of size N that contains one infectious individual. (Note that in Ferguson et al. (1997a), $R_0^{(H)}$ was defined as the number of secondary *cases* generated by a single primary *case*. Both definitions will give identical values of $R_0^{(H)}$ if the birth rate and β_H are constant through time, but not otherwise.) For the Markovian model (4.5), it is trivial to show that $R_0^{(H)} = \beta_H/\sigma$.

4.3 Introducing demography

Let us consider a population with a birth rate $B(t)$ at time t, where the probability of surviving to age a (given birth at time t) is $S(a,t)$, and now assume that individuals of all ages mix randomly. It becomes necessary to integrate over all ages of infectious individuals to obtain the infection hazard:

$$Q(t) = \int_0^\infty \frac{n(t,a')}{N(t)} \int_0^{a'} \beta_H \Psi_H(\tau)\rho(a' - \tau|t - a')d\tau da'. \quad (4.6)$$

Here $n(t,a')$ is the density of hosts of age a' at time t, given by $n(t,a') = B(t-a')S(a',t-a')$ and $N(t) = \int_0^\infty n(t,a')da'$, the total population size at time t.

This form of infection hazard assumes *mass-action* mixing (De Jong et al., 1995), namely that the infection hazard experienced by any host is proportional to the *proportion* of all contacts that are with infectious animals, and that animals of all ages mix homogeneously. The latter assumption means that the infection hazard experienced by any individual is proportional to the fraction of the population that is infectious. This explains the origin of the $N(t)$ denominator in (4.6).

For later ease of notation we define the PDF for the age, a', of animals alive at time t to be

$$\sigma_H(t,a') = \frac{n(t,a')}{N(t)}, \quad (4.7)$$

so

$$Q(t) = \int_0^\infty \sigma_H(t,a') \int_0^{a'} \beta_H \Psi_H(\tau) \rho(a' - \tau | t - a') d\tau da'. \quad (4.8)$$

Note that here and below, we have assumed time-independent survivorship, it being trivial to allow for time-varying survivorship.

The expression for the basic reproduction number, $R_0^{(H)}$, now becomes more complex, as we have to average over all possible ages of the primary infection, and allow for the survivorship of the primary infection during disease incubation:

$$R_0^{(H)}(t) = \int_0^\infty \int_0^\infty q_H(t,a) \nu_H(t,a,\tau) N(t+\tau) d\tau da.$$

Here $q_H(t,a)$ is the distribution of ages of the primary infection, just given by $q_H(t,a) = \sigma_H(t,a)$ here. The mean infectiousness (allowing for survivorship) of the primary infection at time τ after infection at age a, $\nu_H(t,a,\tau)$, is given by

$$\nu_H(t,a,\tau) = \frac{1}{S(a,t-a)} S(a+\tau, t-a) \frac{\beta_H \Psi_H(\tau)}{N(t+\tau)}. \quad (4.9)$$

The first term in this expression just conditions on survival to the age of infection of the primary case, while the second represents the probability of surviving for time τ following infection. The third term is the infectiousness of a single individual in a susceptible population of size N at time τ following infection (note that under the mass-action principle, it is the *proportion* of infectious individuals that determines the infection hazard of a susceptible). Hence

$$R_0^{(H)}(t) = \int_0^\infty \int_0^\infty \frac{S(a+\tau, t-a)}{S(a,t-a)} \beta_H \Psi_H(\tau) \sigma_H(t,a) d\tau da$$

where t is the time of infection of the primary individual. Note that $R_0^{(H)}$ is constant in time only if the birth rate is constant.

Again, it may be informative for readers with a background in epidemic models (Bailey, 1975), but not in survival analysis, to understand the relationship between this model and differential equation models. We define $x(t,a)$ to be the density of susceptible individuals of age a at time t:

$$x(t,a) = n(t,a) \left[1 - p_I(a|t-a)\right].$$

The comparable density of infectious hosts is given by

$$y(t,a) = n(t,a) \int_0^a \Psi_H(\tau)\rho(a - \tau|t - a)d\tau$$

so that

$$Q(t) = \frac{\beta_H}{N(t)} \int_0^a y(t,a')da'.$$

If $\Psi_H(\tau) = \exp(-\eta\tau)$, it is then easy to show from (4.1) and (4.8) that

$$\frac{\partial x}{\partial t} + \frac{\partial x}{\partial a} = -\frac{\beta_H}{N(t)}x \int_0^\infty y(t,a)da - \mu(t,a)x$$

$$\frac{\partial y}{\partial t} + \frac{\partial y}{\partial a} = \frac{\beta_H}{N(t)}x \int_0^\infty y(t,a)da - [\eta + \mu(t,a)]y$$

where

$$\mu(t,a) = -\frac{1}{n(t,a)}\left(\frac{\partial\sigma_H(t,a)}{\partial t} + \frac{\partial\sigma_H(t,a)}{\partial a}\right)$$

and we have used the result

$$\frac{d}{dt} = \frac{\partial}{\partial t} + \frac{da}{dt}\frac{\partial}{\partial a} = \frac{\partial}{\partial t} + \frac{\partial}{\partial a}.$$

The boundary conditions on x and y are $x(t,0) = B(t)$ and $y(t,0) = 0$. Note that it is possible to derive partial differential equations for this system without assuming exponentially increasing infectiousness, but it is necessary to further stratify the infectious population density, y, by τ, the time since infection. We do not pursue that approach here.

If the birth rate is constant $\mu(t,a)$ is independent of t, but in addition, survivorship must be exponential $(S(a) = \exp(-\mu a))$ for $\mu(t,a)$ to be a constant. In this latter case, we can integrate over age, defining $X(t) = \int_0^\infty x(t,a)da$, and $Y(t) = \int_0^\infty y(t,a)da$, and obtain

$$\frac{dX}{dt} = -\beta_H\frac{XY}{N} - \mu X$$

$$\frac{dY}{dt} = \beta_H\frac{XY}{N} - (\eta + \mu)Y.$$

4.4 Age-dependent susceptibility/exposure

We can now extend the model further to include the possibility that susceptibility or exposure varies with age by making the infection hazard age-dependent. In the simplest case, where we assume that mixing is still independent of age, this gives

$$Q(t, a) = g_H(a) \int_0^\infty \sigma_H(t, a') \int_0^{a'} \beta_H \Psi_H(\tau) \rho(a' - \tau | t - a') d\tau da'$$

where $g_H(a)$ represents the susceptibility/exposure of an individual at age a to direct horizontal infection. Since $g_H(a)$ needs to be standardized in some manner, we normalize it to be a PDF on the interval $[0, a_{\max}]$. The function $g_H(a)$ then represents the age at infection distribution for a time-independent infinitesimal force of infection.

Given such age-dependence, it is useful to retain an expression for the force of infection alone:

$$r_H(t) \quad = \quad \frac{Q(t, a)}{g_H(a)} \tag{4.10}$$

$$= \quad \int_0^\infty \sigma_H(t, a') \int_0^{a'} \beta_H \Psi_H(\tau) \rho(a' - \tau | t - a') d\tau da'.$$

Calculating $R_0^{(H)}$ now becomes more complex still. The expected infectivity of the primary infection, $\nu_H(a, \tau)$, remains the same, but its age distribution, $q_H(t, a)$, now depends on $g_H(a)$:

$$q_H(t, a) = \frac{n(t, a) g_H(a)}{\int_0^\infty n(t, a') g_H(a') da'}.$$

We also now have to average over all possible ages of the secondary infections. This gives

$$R_0^{(H)}(t) = \int_0^\infty \int_0^\infty \int_0^\infty q_H(t, a) \nu_H(t, a, \tau) g_H(a') n(t+\tau, a') da' d\tau da \tag{4.11}$$

where $t + \tau$ is the time at which the secondary infection occurs in an animal of age a'. This expression can be simplified thus:

$$R_0^{(H)}(t) = \int_0^\infty \int_0^\infty \frac{W_H(t + \tau)}{W_H(t)} g_H(a) \sigma_H(t+\tau, a+\tau) \beta_H \Psi_H(\tau) d\tau da,$$

where $W_H(t)$ is defined as

$$W_H(t) = \int_0^\infty g_H(a') n(t, a') da'.$$

When the birth rate and survivorship are both time invariant, this reduces to

$$R_0^{(H)}(t) = \int_0^\infty \int_0^\infty g_H(a)\sigma_H(t+\tau, a+\tau)\beta_H \Psi_H(\tau)d\tau da.$$

4.5 The inclusion of different transmission routes

Thus far we have only considered the case of a disease transmitted solely horizontally, through direct contact. In the case of BSE, there are three potential transmission routes: indirect horizontal (through re-cycling in MBM feed), maternal (dam to calf), and direct horizontal. The infection hazard then becomes

$$Q(t,a) = \sum_{j=F,M,H} Q_j(t,a) = \sum_{j=F,M,H} g_j(a)r_j(t)$$

where $g_j(a)$ is the susceptibility/exposure of an individual at age a to infection via route j, and $r_j(t)$ is the force of infection (for unit susceptibility) via transmission route j at time t, conditional on the other routes not having arisen.

The form of $g_j(a)$ and $r_j(t)$ will clearly differ between transmission routes. For maternal transmission, $g_M(a) = \delta(a)$, the Dirac delta function*; i.e. there is an atom of probability of maternal transmission $\beta_M r_M(t)$ at the time of birth. While it is convenient to make use of the integrated hazard when considering both continuous infection risks (horizontal and feed-borne transmission) and maternal transmission, this is only meaningful if the 'product integral' measure is used (Cox and Oakes, 1984, page 15). For our purposes, this can be defined most simply by its operation. As a simple example, let $h(t)$ be a hazard composed of a continuous part, $h_1(t)$, and a discrete element, $h_2\delta(t-t_1)$, at $t = t_1$. The probability of remaining uninfected up to time t is given by the integrated hazard $F(t) = \exp[-\int_0^t h(t')dt']$, which (in the case of mixed continuous and discrete hazards) is defined to be:

$$\begin{aligned}
F(t) &= \exp\left[-\int_0^t h(t')dt'\right] \\
&\equiv \begin{cases} \exp\left[-\int_0^t h_1(t')dt'\right] & t \le t_1 \\ (1-h_2)\exp\left[-\int_0^t h_1(t')dt'\right] & t > t_1 \end{cases}
\end{aligned} \tag{4.12}$$

* It can be viewed as the limit of any continuous PDF (with zero mean) as the variance tends to zero.

Now the PDF for infection at time t, $f(t)$, is given by $f(t) = h(t)F(t)$, with $\int_0^t f(t')dt' = 1 - F(t)$, the CDF for infection by time t. It will prove useful later, however, to have separate expressions for the CDF of infection by time t due to the discrete and continuous hazards:

$$\int_0^t h_1(t)F(t')dt' = \begin{cases} 1 - \exp\left[-\int_0^t h_1(t')dt'\right] & t \le t_1 \\ 1 - h_2 \exp\left[-\int_0^{t_1} h_1(t')dt'\right] & \\ \quad -(1 - h_2)\exp\left[-\int_0^t h_1(t')dt'\right] & t > t_1 \end{cases},$$

(4.13)

$$\int_0^t h_2\delta(t' - t_1)F(t')dt' = \begin{cases} 0 & t \le t_1 \\ h_2 \exp\left[-\int_0^{t_1} h_1(t')dt'\right] & t > t_1 \end{cases}.$$

(4.14)

This formalism enables all transmission routes to be treated in a consistent manner, allowing $r_M(t)$ to be expressed in a very similar form to that for direct horizontal transmission given in (4.10):

$$r_M(t) = \int_0^\infty \sigma_M(t, a') \int_0^{a'} \beta_M \Psi_M(\tau)\rho(a' - \tau | t - a')d\tau da'.$$

Here, however, $\Psi_M(\tau)$ represents the relative infectiousness of a dam through the maternal transmission route at time τ after infection, β_M the maternal transmission probability given maximum infectiousness, and $\sigma_M(t, a)$, the proportion of dams at time t that are of age a. We are assuming that only animals over 2 years of age calve, and that the only animals surviving to that age are female. Then

$$\sigma_M(t, a) = \begin{cases} 0 & a \le 2 \\ \frac{B(t-a)S(a)}{\int_2^\infty B(t-a)S(a)da} & a > 2 \end{cases}.$$

(4.15)

Note that we have neglected the effects of disease-induced mortality. This is reasonable for any population-level model, as the peak cumulative disease incidence in any one cohort never exceeded 2%. However, due to the clustering of cases in herds, this approximation is less valid for herd-level models, though correcting for disease mortality is not as simple as correcting the denominator in (4.7) and (4.15) above, since farmers would be expected to replace stock lost through disease.

Let us now consider the feed-borne transmission route. Here it is infected animals slaughtered prior to disease onset that are fed back to susceptible animals in the form of MBM supplements to cause infection. Therefore, $r_F(t)$ has the following form:

$$r_F(t) = \int_0^\infty \sigma_F(t,a') \int_0^{a'} \beta_F(t)\Psi_F(\tau)\rho(a'-\tau|t-a')d\tau da' \quad (4.16)$$

where $\sigma_F(t,a)$ is the proportion of animals slaughtered at time t that were of age a:

$$\sigma_F(t,a) = \frac{B(t-a)\frac{dS}{da}}{\int_0^\infty B(t-a)\frac{dS}{da}da}.$$

Note that we have explicitly allowed transmission coefficient, β_F, for feed-borne transmission to depend upon time t. In the case of the maternal and horizontal routes, we have assumed up to now that transmission probabilities are independent of time, since these transmission routes are not subject to change in the same way as changes in rendering practices, together with controls on the use of bovine MBM in animal feed, might affect the probability of infected material being recycled through the bovine food chain. However, in the expressions presented below for arbitrary transmission route j, we allow, for generality, the transmission coefficient, $\beta_j(t)$, to vary with time.

Lastly, the formalism above contains one major simplifying assumption − that the infectiousness of an infected animal does not depend on the route through which it was infected. If this assumption is relaxed, it becomes necessary to keep track of infection incidence through each transmission route separately. Denoting the PDF for infection at age a through route j given birth at time t_0 to be $\rho_j(a|t_0)$, $r_j(t)$ is given by

$$\begin{aligned} r_j(t) &= \int_0^\infty \sigma_j(t,a') \int_0^{a'} \beta_j(t) \\ &\times \sum_{j'=F,M,H} [\Psi_{jj'}(\tau)\rho_{j'}(a'-\tau|t-a')]\,d\tau da' \quad (4.17) \end{aligned}$$

where $\Psi_{jj'}(\tau)$ is the relative infectiousness via route j of an animal that was infected via route j', at time τ after infection.

It is easy to show that $\rho_j(a|t_0)$ is given by:

$$\rho_j(a|t_0) = Q_j(t_0+a,a)\exp\left[-\int_0^a \sum_{k=F,M,H} Q_k(t_0+a',a')da'\right].$$

4.6 R_0 for systems with multiple transmission routes

What is a typical primary infection in the case of a disease with multiple transmission routes? Also, what is a typical secondary infection? These are relevant questions, as age-at-infection, infectivity and susceptibility might differ for different transmission routes. We therefore need some way to average over the contribution of all transmission routes, while preserving the threshold property for epidemic growth of $R_0 > 1$.

Let us define $R_0^{(jk)}$ to be the expected number of secondary infections via route k generated by a single primary infection occurring via route j in an entirely susceptible population. Generalizing the earlier expression (4.11) gives

$$R_0^{(jk)}(t) = \int_0^\infty \int_0^\infty \int_0^\infty q_j(t,a)\nu_k(t,a,\tau)g_k(a')n(t+\tau,a')da'\,d\tau\,da \tag{4.18}$$

where $q_j(t,a)$, as before, is defined by

$$q_j(t,a) = \frac{n(t,a)g_j(a)}{\int_0^\infty n(t,a')g_j(a')da'},$$

and $\nu_k(t,a,\tau)$ is given by an expression analogous to that for $\nu_H(a,\tau)$ (4.9), namely

$$\nu_k(t,a,\tau) = \frac{\sigma_k(t+\tau,a+\tau)}{n(t,a)}\beta_k(t+\tau)\Psi_k(\tau).$$

The equation (4.18) can then be simplified as

$$R_0^{(jk)}(t) = \int_0^\infty \int_0^\infty \frac{W_k(t+\tau)}{W_j(t)}g_i(a)\sigma_k(t+\tau,a+\tau)\beta_k(t+\tau)\Psi_k(\tau)d\tau\,da,$$

where $W_j(t)$ is defined as

$$W_j(t) = \int_0^\infty g_j(a')n(t,a')da'.$$

If d_k defines the probability of the primary case being infected through transmission route k, then the matrix product

$$d'_j = \sum_k R_0^{(jk)}d_k$$

defines the distribution of 'next-generation' infection types. Therefore, $R_0^{(jk)}$ is termed a *next-generation operator*, and it can be shown that the overall R_0 of the system is given by the dominant eigenvalue of this operator (Heesterbeek and Dietz, 1996) –

i.e. when the dominant eigenvalue is greater than 1, initial growth of any epidemic will be exponential.

It is straightforward to extend this analysis to the case where infectiousness through a particular route depends on the route through which the infectious animal was infected itself — just substitute $\nu_{jk}(a,\tau)$ for $\nu_k(a,\tau)$ in (4.18), where $\nu_{jk}(t,a,\tau)$ is obtained by substituting $\Psi_{kj}(\tau)$ (see equation 4.17) for $\Psi_k(\tau)$ in the definitions of $\nu_k(t,a,\tau)$.

Up to this point, all expressions for $R_0(t)$ have defined t to be the time of infection of the primary infection. However, in the case of BSE, most infections were through the feed-borne route, with secondary infections occurring on the death of the primary infection. It is therefore more useful to define R_0 relative to the time of death of the primary infection, t_s, as this more temporally correlated with the secondary infection process:

$$R_0^{(jF)}(t_s) = \beta_F(t_s) \int_0^\infty \int_0^\infty \frac{W_F(t_s)}{W_i(t_s - \tau)} g_j(a)\sigma_F(t_s, a+\tau)\Psi_F(\tau)d\tau da.$$
(4.19)

It is more complex to change time coordinates in the case of maternal or direct horizontal secondary infections, however, and so we omit the details here.

Lastly, it will prove informative to examine the average number of feed-borne secondary infections caused (in an entirely susceptible population) by one maximally infectious (*i.e.* $\Psi_F(\tau) = 1$) animal that is slaughtered prior to disease onset, $I^{(FF)}(t_s)$:

$$I^{(FF)}(t_s) = \beta_F(t_s) \int_0^\infty \int_0^\infty \frac{W_F(t_s)}{W_F(t_s - \tau)} g_F(a)\sigma_F(t_s, a + \tau)d\tau da.$$
(4.20)

4.7 Solving the model

We have been left with the following integro-differential equation for $p_I(a|t_0)$:

$$\frac{\partial p_I(a|t_0)}{\partial a} = [1 - p_I(a|t_0)]$$

$$\times \int_0^\infty \int_0^{a'} \left[\sum_{j=F,M,H} g_j(a)\sigma_j(t,a')\beta_j(t)\Psi_j(t,\tau) \right]$$

$$\times \frac{\partial p_I}{\partial a}(a' - \tau|t - a')d\tau da'. \qquad (4.21)$$

This equation is, in general, analytically intractable, and − more importantly from our point of view − very computationally intensive even to solve numerically. Let us step back, therefore, and examine to what extent the system can be computationally simplified.

The key point is that in using back-calculation, we do not need to explicitly model the mechanism of each transmission process. In the absence of maternal and horizontal transmission, our purpose is to obtain an estimate of the time-varying force of infection $r_F(t)$. Unless we are explicitly interested in the form of the infectivity function $\beta_F(t)\Psi_F(t,\tau)$, we can directly estimate $r_F(t)$ non-parametrically using (from (4.2))

$$\rho(a|t_0) = Q_F(t_0 + a, a)\exp\left[-\int_0^a Q_F(t_0 + a', a')da'\right]$$

$$= g_F(a)r_F(t_0 + a)\exp\left[-\int_0^a g_F(a')r_F(t_0 + a')da'\right].$$

Given an estimate of $r_F(t)$, we are then free to explore the relationship between the infectiousness model and the resulting time-dependent transmission probability profile, $\beta_F(t)$ − representing the effectiveness of the rendering process at deactivating infectious material − by using (4.16) to estimate $\beta_F(t)$:

$$\beta_F(t) = r_F(t) \bigg/ \left\{ \int_0^\infty \sigma_F(t,a) \int_0^a \Psi_F(\tau)g_F(a - \tau)r_F(t - \tau) \right.$$

$$\left. \times \exp\left[-\int_0^{a-\tau} g_F(a')r_F(t - a + a')da'\right] d\tau da \right\}.$$

This illustrates an important point: in the absence of specific information on how transmission probabilities vary through the duration of an epidemic, it is impossible to distinguish between different infectiousness $(\Psi_F(\tau))$ models on the basis of goodness-of-fit estimates alone: even if we directly estimate $\beta_F(t)$ (without restriction) by solving (4.21) in the back-calculation procedure, rather than directly estimate $r_F(t)$, the resulting $r_F(t)$ estimate — and goodness-of-fit — would be independent of the form of $\Psi_F(\tau)$ assumed. As we shall see later, however, there are epidemiological criteria on which to make judgements of alternative infectiousness models.

When it comes to maternal and direct horizontal transmission, however, we no longer have the luxury of an unknown and time-varying transmission coefficient. As β_M and β_H are largely determined by disease pathogenesis and the rate of between-animal contact, the only reasonable assumption is that neither varied in value during the epidemic. How then do we avoid the problem of having to explicitly solve (4.21)? The answer is through the use of a *generation* expansion; *i.e.* to calculate $r_M(t)$ and $r_H(t)$ accurate up to a certain number of generations of transmission through each route. Clearly, this is only possible if the chain of infection via non feed-borne routes caused by one primary infection is finite — meaning that the basic reproduction number, $R_0^{(j)}$ for each transmission route must be below 1. In the case of maternal transmission, this is always satisfied (assuming a stable or declining population size), as the probability that a dam infects its calf is at most 1. For direct horizontal transmission this need not be the case, but a generation expansion of $r_H(t)$ will at least allow easy exploration of the $R_0^{(H)} < 1$ case.

In the case where only one of the maternal and horizontal transmission routes is present, the force of infection due to that route is defined by:

$$r_j(t) = \beta_j \int_0^\infty \sigma_j(t,a') \int_0^{a'} \Psi(\tau) \left[\sum_{k=F,j} g_k(a'-\tau) r_k(t-\tau) \right]$$

$$\times \exp \left[-\int_0^{a'-\tau} \sum_{k=F,j} g_k(a'') r_k(t-a'+a'') da'' \right] d\tau da'$$

$$(4.22)$$

where $j = M, H$ and so β_j is independent of t. We can represent this more concisely (and simplify future expressions) using integral operators. Define $\mathcal{F}_j$ and $\mathcal{G}_j$ thus:

$$\mathcal{F}_j y = \int_0^\infty \sigma_j(t, a') \int_0^{a'} \Psi(\tau) g_j(a' - \tau) y(t - \tau)$$
$$\times \exp\left\{ -\int_0^{a'-\tau} [g_F(a'') r_F(t - a' + a'') \right.$$
$$\left. + g_j(a'') y(t - a' + a'')]\, da'' \right\} d\tau\, da'$$

$$\mathcal{G}_j y = \int_0^\infty \sigma_j(t, a') \int_0^{a'} \Psi(\tau) g_F(a' - \tau) r_F(t - \tau)$$
$$\times \exp\left(-\int_0^{a'-\tau} \{g_F(a'') r_F(t - a' + a'') \right.$$
$$\left. + g_j(a'')[1 - y(t - a' + a'')]\} \, da'' \right) d\tau\, da'$$

where $y(t)$ is an arbitrary continuous function of t. To simplify subsequent expressions, $\mathcal{G}_j y$ is expressed in terms of $1 - y$. Then (4.22) can be written

$$r_j(t) = \beta_j \left[\mathcal{G}_j(1 - r_j) + \mathcal{F}_j r_j \right].$$

Let $r_j^{(n)}(t)$ be a representation of $r_j(t)$ accurate to generation n. Then $r_j^{(1)}$ — which assumes that all 'zeroth' generation infections were feed-borne — is given by

$$r_j^{(1)}(t) = \beta_j \mathcal{G}_j(1 - r_j^{(0)}) = \beta_j \mathcal{G}_j 1,$$

and $r_j^{(n)}(t)$ can then be generated recursively thus

$$r_j^{(n)} = \beta_j \left[\mathcal{F}_j r_j^{(n-1)} + \mathcal{G}_j \left(1 - r_j^{(n-1)} \right) \right]. \qquad (4.23)$$

$\mathcal{F}_j$ and $\mathcal{G}_j$ are *next-generation operators* of a rather complex and non-linear form; $\beta_j \mathcal{F}_j$ calculates the proportion infected in one generation through transmission route j by animals who were themselves infected through route j, while $\beta_j \mathcal{G}_j$ calculates the proportion infected through route j by animals who were feed-infected.

In practice, $\mathcal{F}_j$ and $\mathcal{G}_j$ are useful in numerical evaluation of hazards due to maternal or horizontal transmission, but their complexity makes them of relatively little analytical use (say in proving convergence of (4.23)), except in the case of maternal transmission, where use of generalized version of the identities (4.12), (4.13) and (4.14) results in dramatic simplification:

$$\mathcal{F}_M y = \int_0^\infty \sigma_j(t,a')\Psi(a')y(t-a')da'$$

$$\mathcal{G}_M y = \int_0^\infty \sigma_j(t,a') \left\{ \int_0^{a'} \Psi(\tau)g_F(a'-\tau)r_F(t-\tau) \right.$$
$$\left. \times \exp[-\int_0^{a'-\tau} g_F(a'')r_F(t-a'+a'')da'']d\tau \right\} y(t-a')da'.$$

Then, using the linearity of $\mathcal{F}_M$ and $\mathcal{G}_M$, (4.23) can be rewritten thus:

$$r_M^{(n)} = \beta_M \left[(\mathcal{F}_M - \mathcal{G}_M)r_M^{(n-1)} - \mathcal{G}_M 1 \right].$$

Hence,

$$r_M^{(n)} = \left[\sum_{i=1}^n \beta_M^{i-1}(\mathcal{F}_M - \mathcal{G}_M)^{i-1} \right] \beta_M \mathcal{G}_M 1 \qquad (4.24)$$

where $\lim_{n\to\infty} r_M^{(n)} = r_M$ so long as $\beta_M < 1$.

Note that the generation expansion of $r_M(t)$ is equivalent to a power-series expansion in β_M only because $\mathcal{F}_M$ and $\mathcal{G}_M$ are linear operators — which in turn is due to the discrete nature of the maternal infection hazard.

In numerical calculations of $r_M(t)$ and $r_H(t)$, it proves simplest to iterate the recursion relation (4.23).

4.8 Heterogeneity in susceptibility

Until now, we have assumed that all individuals in the cattle population are equally susceptible to infection. As many other TSEs show genetically determined variable susceptibility, it is worth extending the model to allow for multiple susceptibility classes. Let p_g denote the proportion of the total population in class g and let s_i denote the relative susceptibility of class i. Thus, assuming the relative susceptibilities are independent of infection risk, $Q_g(t,a)$, the infection hazard for class g, is defined by

$$Q_g(t,a) = s_g Q(t,a)$$

where $s_1 = 1$. We therefore get a different PDF for the age at infection for each susceptibility class, g, via each transmission route, j:

$$\rho_{jg}(a|t_0) \;=\; s_g g_j(a) r_j(t_0 + a)$$

$$\times \exp\left[-s_g \int_0^a \sum_{k=F,M,H} g_k(a') r_k(t_0 + a') da' \right].$$

Similarly, each susceptibility class, g, gives rise to a different force of infection so the total force of infection through transmission route j is now

$$r_j(t) \;=\; \beta_j(t) \int_0^\infty \sigma_j(t, a') \int_0^{a'} \Psi_j(\tau)$$

$$\times \sum_g \sum_{k=F,M,H} p_g \rho_{kg}(a' - \tau | t - a') d\tau da'.$$

However, this formulation assumes that the susceptibility class of an infectious animal is independent of that of any animal it infects, which is clearly false in the case of maternal transmission. More generally, therefore, we define a force of infection arising from animals in class g, through transmission route j, as

$$r_{jg}(t) \;=\; \beta_j(t) \int_0^\infty \sigma_j(t, a') \int_0^{a'} \Psi_j(\tau)$$

$$\times \sum_{k=F,M,H} \rho_{kg}(a' - \tau | t - a') d\tau da'$$

We then obtain

$$\rho_{jg}(a|t_0) \;=\; s_g \sum_{g'} \frac{p_{jg|g'} p_{g'}}{p_g} g_j(a) r_{jg'}(t_0 + a)$$

$$\times \exp\left[-s_g \int_0^a \sum_{k=F,M,H} g_k(a') r_{kg'}(t_0 + a') da' \right]$$

where $p_{jg|g'}$ is the proportion of infectious contacts via transmission route j that an animal in susceptibility class g makes with an animal in susceptibility class g'. Clearly, for random mixing, $p_{Fg|g'} = p_{Hg|g'} = p_g$. However, $p_{Mg|g'}$ represents the probability that a dam of type g' will have a calf of type g.

4.9 Incidence of disease

Discussion of incubation period distribution (IPD) has been avoided until now. However, as it is the incidence of disease onset, rather than infection, that is actually measured in case notification statistics, it is necessary to include the IPD to complete our survival model. If we define $\phi_C(t_0, a)$ to be the PDF of an animal experiencing disease onset at age a, then

$$\phi_C(t_0, a) = S(a) \sum_g p_g \int_0^a \sum_{j=F,M,H} \rho_{jg}(a - u|t_0) f_j(u) du \quad (4.25)$$

where $f_j(u)$ is the PDF for incubation period u for an animal infected through transmission route j.

We also need to allow for possible under-reporting. If $\Lambda(t)$ is the probability that a case that onset at time t is reported, then the PDF for an animal born at time t_0 being reported as a case at age a is given by

$$\phi_{RC}(t_0, a) = \Lambda(t_0 + a) S(a) \sum_g p_g \int_0^a \sum_{j=F,M,H} \rho_{jg}(a - u|t_0) f_j(u) du.$$
$$(4.26)$$

Finally, relative infectiousness may be better described by the time remaining before disease onset, v, rather than by the time since infection, τ. If $\Omega_j(v)$ is the relative infectiousness of an animal infected through transmission route j at time v before disease onset, then $\Psi_j(\tau)$ is given by

$$\Psi_j(\tau) = \int_0^\infty \Omega_j(v) f_j(\tau + v) dv.$$

4.10 Maximum likelihood methods for back-calculation

Assuming all animals are independent, the age-stratified incidence data arise from a multinomial distribution and the likelihood can be written in terms of the PDF $\phi_{RC}(t_0, a)$. Let N_i be the number of calves born in the time interval T_{i-1} to T_i, such that $N_i = \int_{T_{i-1}}^{T_i} B(t) dt$. Let $X_{j,i}$ be the number of cases among calves born in the time interval T_{i-1} to T_i with onset between ages A_{j-1} and A_j, where $A_0 = 0$, and $A_{j_{\max(T_i)}}$ is given by the maximum possible age of onset. The maximum age $A_{j_{\max(T_i)}}$ varies with T_i since cohorts have been observed for variable amounts of time. Thus, $N_i - \sum_{j=1}^{j_{\max(T_i)}} X_{j,i}$ is the number of calves born in time interval

$(T_{i-1}, T_i]$ that were not observed to experience disease onset by age $A_{j_{\max(T_i)}}$. The complete data log likelihood (l) is written as a sum over cohorts, i:

$$l = \sum_i \left[\sum_{j=1}^{j_{\max(T_i)}} X_{j,i} \log \left(\frac{\int_{T_{i-1}}^{T_i} \int_{A_{j-1}}^{A_j} B(t)\phi_{RC}(t,a)da\,dt}{\int_{T_{i-1}}^{T_i} B(t)dt} \right) \right.$$
$$\left. + \left(N_i - \sum_{j=1}^{j_{\max(T_i)}} X_{j,i} \right) \log \left(1 - \frac{\int_{T_{i-1}}^{T_i} \int_0^{A_{j_{\max(T_i)}}} B(t)\phi_{RC}(t,a)da\,dt}{\int_{T_{i-1}}^{T_i} B(t)dt} \right) \right]$$

ignoring additive constants.

Maximization of this likelihood has to utilize non-linear optimization techniques — $e.g.$ direction-set (Powell's) methods, simulated annealing (Press $et\ al.$, 1992) or sub-energy tunnelling methods such as TRUST (Barhen $et\ al.$, 1997). In high-dimensional parameter spaces, these algorithms are very computationally intensive, often requiring the sampling of hundreds of thousands of parameter points to reach the maximum likelihood point. Furthermore, when such an algorithm converges on a particular local maximum, there is no guarantee that the point reached is the global maximum within the finite hypercube of parameter space within which one has performed the search (unconstrained searches are even more problematic). Thus it is necessary to perform the maximization multiple times from different starting points, and (optimally) with different algorithms.

4.11 Model goodness-of-fit

The goodness-of-fit of the model can be judged by comparing the maximum model likelihood with the saturated data likelihood using a likelihood ratio statistic.

The likelihood analyses assume conditional independence of the observations. If overdispersion has been ignored, the nominal standard errors for multinomial models are underestimates. Thus, the standard errors need to be inflated and the goodness-of-fit statistic needs to be correspondingly deflated. This is particularly relevant in the case of the BSE epidemic, due to the observed clustering of cases (Chapter 8). This phenomenon results from the reduction in the effective sample size caused by correlation and would affect our standard error and goodness-of-fit measures as well. Models that explicitly incorporate herd-level clustering are discussed in Chapters 8 and 9.

Conversely, however, when one samples the distribution of likelihood ratio goodness-of-fit statistics directly from the model using bootstrap techniques, the effective degrees of freedom appear to be consistently fewer than the number obtained by subtracting the number of model parameters and constraints from the number of multinomial outcomes. This appears to arise because of the relatively large number of outcomes with low probabilities. This effect will tend to counterbalance the former, in that it will result in over-optimistic estimates of the goodness-of-fit. However, these effects will not affect our qualitative conclusions.

4.12 Confidence and prediction intervals

The simultaneous likelihood ratio confidence region for all of the maximum likelihood estimated parameters contains all combinations of parameters that provide a similar goodness-of-fit to the observed data, as measured by the likelihood ratio statistic. Thus, the likelihood ratio 95% confidence regions are defined by the multidimensional region containing only the combinations of parameters corresponding to a log likelihood within $\frac{1}{2}\chi^2_{P,.95}$ of the maximum log likelihood where P is the number of parameters estimated using maximum likelihood methods.

The corresponding confidence interval for any resulting quantity is obtained from the range of values arising from the combinations of parameters contained in the confidence region. Prediction intervals can be similarly defined to bound the predictions resulting from all combinations of parameters providing a similar goodness-of-fit to the observed data.

The high dimensionality and complex geometry of the parameter space make the characterization of this region highly computationally intensive. It is therefore simplest to restrict the search to determining the boundaries of the region that intersect with the individual parameter axes through the best fit point; a further refinement of this procedure is to randomly sample vectors from the best fit point and determine their intersection with the confidence region boundary. However, using this type (or any type − e.g. including the points sampled during the likelihood maximization procedure) of sampling of parameter space inevitably causes the widths of confidence limits thus calculated to be lower bounds, especially in the case of highly correlated parameters (e.g. those defining the age-dependent susceptibility/exposure and incubation period distributions). It is therefore critical to perform sensitivity analyses that are explicitly aimed at identifying correlated parameters. Moreover, through the use of many different functional forms for all parametric distributions, it is possible to start to explore the space of possible models, as well as the parameter space for individual models.

4.13 Conclusion

In this chapter we have developed the epidemic models necessary to model the transmission dynamics of BSE at the population-wide level (*i.e.* not allowing for herd-level heterogeneity). These models were developed within a survival analysis framework by deriving mechanistic expressions for the infection hazard, though the equivalence between this approach and more traditional compartmental differential equation based epidemic models has been highlighted. A variety of techniques to simplify model evaluation and parameter estimation were also outlined. In the next chapter we turn to the practical application of the models discussed here, and present estimates of the past pattern of infection, changes in the basic reproduction number of R_0 over time, and an extensive analysis of the sensitivity of model results to variation in the assumptions made regarding key distributions and parameters.

CHAPTER 5

Population models: results and sensitivity analyses

5.1 Introduction

In this chapter, we summarize the results of back-calculation analyses of the BSE epidemics in Great Britain (Ferguson *et al.*, 1997a) and Northern Ireland (Ferguson *et al.*, 1998). The results provide a basis for the analysis of the vCJD epidemic (Chapter 10) by providing estimates of the numbers of infected animals consumed through time, stratified by incubation stage. In addition, estimates of the time-dependent feed risk, and corresponding basic reproduction number, $R_0^{(FF)}$, provide information on the effectiveness of the ruminant feed bans introduced in July 1988 in Great Britain and in January 1989 in Northern Ireland. Also of great interest to policy makers and the public are the predictions of future BSE cases.

We characterize the sensitivity of these results to the assumptions about:

- The functional form of the incubation period and age-dependent susceptibility/exposure distributions

- The profile of cattle-to-cattle infectivity as a function of incubation stage

- The inclusion of non-reported case carcasses in cattle feed

- The level of maternal transmission and the duration of maternal infectiousness

- The level of horizontal transmission and the duration of horizontal infectiousness

- Under-reporting of cases in the early stages of the epidemic

Earlier work also examined the sensitivity of results to genetically determined susceptibility to infection, the resampling performed to address the biases in the reported age-at-onset data (described in Section 3.2.1), the form of the feed-risk profile through time, and demographic factors including birth rate, birth seasonality and survivorship (Ferguson *et al.*, 1997a).

Although it would involve too much computing to examine every combination of factors, the key results (including the estimated number of infected animals slaughtered for consumption and future case predictions) are shown to be largely robust to changes in the parametric forms of key distributions. Other factors, such as the levels of maternal and horizontal transmission, affect the results in a predictable manner.

We take a model with no maternal or horizontal transmission as our baseline for comparison. Animals are assumed to exhibit no genetically variable susceptibility to infection with the aetiological agent of BSE. The feed-risk profile is modelled using a spline with 20 knots, with geometric interpolation between adjacent knots, and under-reporting is fitted prior to July 1988. Where maternal or horizontal transmission is incorporated into the model, it is assumed that the incubation period distribution does not depend on the route of infection (*i.e.* $f_F(u) = f_M(u) = f_H(u) = f(u)$). The main body of results for Great Britain presented here relate to the resampled incidence data (see Section 3.2.1) available through mid-1996 as originally presented in Ferguson *et al.* (1997a), while all of the results for Northern Ireland presented here relate to the raw incidence data available through mid-1997 as originally presented in Ferguson *et al.* (1998).

All of the analyses described here utilize incidence data stratified on a yearly basis, though calvings were calculated on a quarterly basis using the National Milk Records data (Section 3.3; Donnelly *et al.*, 1997b). Serious biases in the cases database, described in Section 3.2.1, and the lack of birth seasonality data prior to 1986 impair our ability to obtain meaningful results from more finely stratified incidence data.

5.2 Distributional forms

A wide range of parametric forms were explored for the incubation period and age-dependent susceptibility/exposure distributions, examples of which are given in Tables 5.1 and 5.2. Incubation period distributions A and B are Gamma and Weibull distributions, respectively, with mean α_2 and an explicitly added time delay of $(1 - \alpha_1)\alpha_2$. Incubation period distribution C arises from a mechanistic model of disease pathogenesis proposed by Medley and

Table 5.1 *Incubation period distributions, $f(u)$. Parameters are α_1, α_2 and α_3, and all are restricted to be positive. Functional forms given are un-normalized.*

Code	Functional form	
A	$0,$ $(u - (1 - \alpha_1)\alpha_2)^{\alpha_1^2 \alpha_2^2 / \alpha_3 - 1}$ $\times \exp\left(-\frac{[u - (1 - \alpha_1)\alpha_2]\alpha_1\alpha_2}{\alpha_3}\right),$	$u \leq (1 - \alpha_1)\alpha_2$ $u > (1 - \alpha_1)\alpha_2$
B	$0,$ $(u - (1 - \alpha_1)\alpha_2)^{\alpha_3 - 1}$ $\times \exp\left[-\left(\frac{(u - (1-\alpha_1)\alpha_2)\Gamma(1+1/\alpha_3)}{\alpha_1\alpha_2}\right)^{\alpha_3}\right],$	$u \leq (1 - \alpha_1)\alpha_2$ $u > (1 - \alpha_1)\alpha_2$
C	$\left(\frac{\alpha_2 \exp(-u/\alpha_1)}{\alpha_3}\right)^{\alpha_2^2/\alpha_3} \exp\left[-\frac{\alpha_2 \exp(-u/\alpha_1)}{\alpha_3}\right]$	

Short (1996). The underlying model assumes that the prion density grows exponentially, at rate γ_1 from an initial dose d_0, causing the onset of clinical signs of disease when the prion density reaches a critical level. Arbitrarily setting this critical level equal to unity and assuming the initial doses arise from the distribution $h(d_0)$, the incubation period conditional on the initial dose is

$$u = -\frac{\log d_0}{\gamma_1}$$

and the distribution of incubation period is given by

$$f(u) = -h\left[d_0(u)\right]\frac{dd_0}{du} = h\left[\exp(-\gamma_1 u)\right]\gamma_1 \exp(-\gamma_1 u).$$

An initial time delay in the incubation period distribution can be obtained if the initial dose distribution peaks at doses far below the critical level. Assuming the initial doses arise from a Gamma distribution gives rise to incubation period distribution C. Nearly identical results were obtained assuming the Beta or Weibull distributions.

In the absence of data on which to base a choice of the parametric form of the age-dependent susceptibility/exposure distribution, we examine a wide range of parametric forms (Table 5.2). Assuming

Table 5.2 *Age-dependent susceptibility/exposure distributions, $g(a)$, for $0 \le a \le 18$. Parameters are γ_1, γ_2, γ_3 and γ_4, and all are restricted to be positive. Functional forms given are un-normalized.*

Code	Functional form
1	$g(a) = \exp(-a/\gamma_1)$
2	$g(a) = \exp(-a/\gamma_1) + \gamma_2$
3	$g(a) = a^{\gamma_2 - 1}\exp(-a/\gamma_1)$
4	$g(a) = \begin{cases} \exp(-a/\gamma_1) + \gamma_2, & a \le 2 \\ 2\left(\exp(-a/\gamma_1) + \gamma_2\right), & a > 2 \end{cases}$
5	$g(a) = \begin{cases} 1, & a \le \gamma_2 \\ \exp(-a/\gamma_1) + \gamma_3, & a > \gamma_2 \end{cases}$
6	$g(a) = 1$
7	$\int_0^a g(a')da' = \{1 - \exp\left[-(\gamma_1 a)^{\gamma_2}\right]\} \times \{1 - \exp\left[-(\gamma_3 a)^{\gamma_2 + \gamma_4}\right]\}$
8	As 7, but with step function exposure, doubling at 2 years of age

constant exposure, distributions 1 and 2 represent exponentially decaying susceptibility, with a constant background susceptibility (2) and no background (1). Similarly assuming constant exposure, distribution 3 represents Gamma-distributed susceptibility with no background. Distribution 4 has the same susceptibility profile as distribution 2 but with step function exposure doubling at 2 years of age, motivated by data indicating that exposure to MBM increases at that age (see Figure 2.7). Assuming constant exposure, distribution 5 represents constant susceptibility to age γ_2 then exponentially decaying susceptibility with a constant background. Distribution 6 represents constant susceptibility and exposure. Distributions 7 and 8 represent empirically derived, extremely flexible susceptibility distributions (Anderson *et al.*, 1996), with constant exposure and step function exposure doubling at 2 years of age, respectively.

Among these distributions, only 7 and 8 are able to produce a

Table 5.3 *Model results for combinations of incubation period and age-dependent susceptibility/exposure distributions fitted to resampled age-stratified incidence data for Great Britain. For each model, the likelihood ratio goodness-of-fit statistic, X^2, and its degrees of freedom (d.f.) are presented along with the estimated incidence of infections 1974 − 1995 and the predicted incidence of cases from 1997−2001. All models assume no maternal or horizontal transmission and allow under-reporting of cases prior to July 1988.*

| $g(a)$ | Incubation period distribution, $f(u)$ | | |
	A	B	C
X^2 (d.f.)	1731 (221)	2238 (221)	1310 (221)
1 Infect. 1974 − 95	945,000	905,000	1,024,000
Cases 1997 − 2001	5,060	4,390	6,960
X^2 (d.f.)	1026 (220)	1612 (220)	800 (220)
2 Infect. 1974 − 95	955,000	920,000	1,117,000
Cases 1997 − 2001	5,580	4,720	18,900
X^2 (d.f.)	469 (220)	1296 (220)	350 (220)
3 Infect. 1974 − 95	1,046,000	1,050,000	1,039,000
Cases 1997 − 2001	18,200	26,200	11,000
X^2 (d.f.)	1053 (220)	1645 (220)	748 (220)
4 Infect. 1974 − 95	1,023,000	928,000	1,110,000
Cases 1997 − 2001	13,600	5,200	19,000
X^2 (d.f.)	1027 (219)	1546 (219)	800 (219)
5 Infect. 1974 − 95	955,000	924,000	1,117,000
Cases 1997 − 2001	5,580	4,860	18,900
X^2 (d.f.)	74,116 (222)	72,990 (222)	74,685 (222)
6 Infect. 1974 − 95	11,144,000	11,103,000	11,129,000
Cases 1997 − 2001	101	114	93
X^2 (d.f.)	377 (218)	710 (218)	308 (218)
7 Infect. 1974 − 95	926,000	931,000	954,000
Cases 1997 − 2001	9,550	16,200	9,340
X^2 (d.f.)	383 (218)	724 (218)	313 (218)
8 Infect. 1974 − 95	927,000	926,000	953,000
Cases 1997 − 2001	9,500	15,600	9,390

Parameter estimates are given in Ferguson *et al.* (1997a).

highly asymmetric form rising sharply to an early sharp peak but dropping more slowly. It should be noted that the model structure does not allow completely general forms of both the incubation period and the age-dependent susceptibility/exposure distribution to be fitted simultaneously. However, sensitivity analysis of the model results is highly informative.

5.3 Model results

Results from the maximum likelihood fits of the back-calculation model to the age-stratified case incidence data for Great Britain for each possible combination of the functional forms presented in Tables 5.1 and 5.2 are presented in Table 5.3. The qualitative effects of the distributional combinations observed in these fits to the resampled incidence data were consistent with those observed in fits to the raw data (Ferguson *et al.*, 1997a).

Incubation period distribution C always provided a better fit to the data than did distributions A and B. The best fitting age-dependent susceptibility/exposure distributions (forms 7, 8 and 3 in that order) share increased probability density in their tails, though the peak occurs at approximately one year of age for form 7 and 8 and at birth for form 3. The assumption of uniform susceptibility/exposure (A6, B6, and C6) is clearly inconsistent with the data.

All distributional combinations, except A6, B6 and C6, yield estimated mean incubation periods between 4.7 and 5.3 years. The 95% confidence interval for the mean incubation period was found to be within the range 4.75 to 5.00 years for the best fitting model for which the mean incubation period can be fixed (A7, since the mean cannot be fixed for incubation period distribution C). The estimated total incidence of infections between 1974 and 1995 increases from 895,000 and 951,000, and predicted case incidence between 1997 and 2001 increases from 8,600 to 10,300 as the mean incubation period increases from 4.75 to 5.00 years. These increasing trends reflect the fact that longer incubation periods mean that infected animals are less likely to survive long enough to experience disease onset. Thus, for a fixed number of cases, more infections are required.

Figure 5.1 presents the maximum likelihood incubation period and age-dependent susceptibility/exposure distributions for model C7. Figures 5.2 and 5.3 give an indication of the quality of the

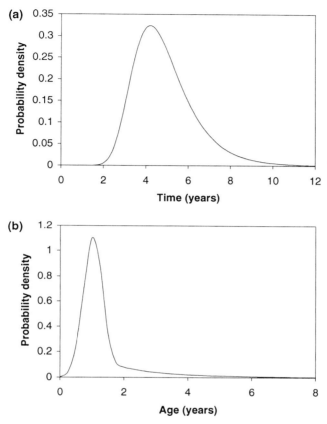

Figure 5.1 *The maximum likelihood estimates of (a) the incubation period distribution of form C and (b) the age-dependent susceptibility/exposure distribution of form 7.*

fit of this model with no maternal or horizontal transmission. The estimated trends in past infection underlying these fitted case incidence profiles (Figure 5.4) illustrate the time lag and the difference in scale — caused respectively by the long mean incubation period and the survivorship of cattle — between the epidemic of infections and the resulting epidemic of cases of BSE. Each 95% confidence interval presented is based on the minimum and maximum value obtained for that function of model parameters over the range of

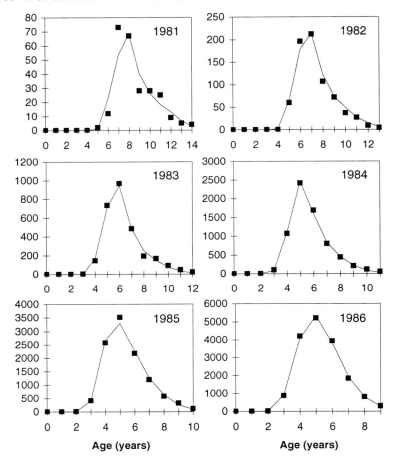

Figure 5.2 *The fitted incidence of reported BSE cases in Great Britain by birth cohort and age at onset for cohorts 1981 to 1986 with the observed case incidence (squares). The model assumed an incubation period distribution of form C and an age-dependent susceptibility/exposure distribution of form 7.*

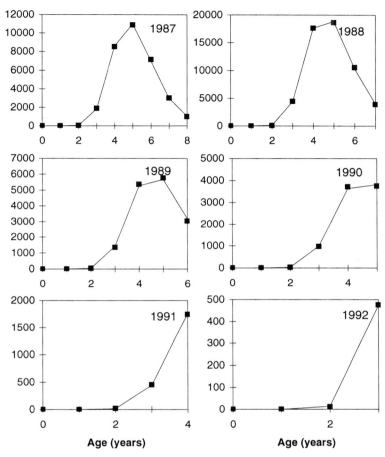

Figure 5.3 *The fitted incidence of reported BSE cases in Great Britain by birth cohort and age at onset for cohorts 1987 to 1992 with the observed case incidence (squares). The model assumed an incubation period distribution of form C and an age-dependent susceptibility/exposure distribution of form 7.*

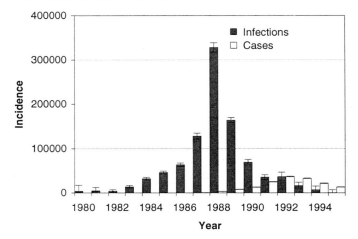

Figure 5.4 *Estimated annual incidence of infections and cases for Great Britain assuming no maternal and no horizontal transmission (with 95% confidence intervals).*

parameter values within the univariate 95% likelihood ratio confidence intervals.

An estimated 467,000 $(460,000 - 482,000)$ infected animals were slaughtered for meat in Great Britain prior to the introduction of the specified bovine offal ban in November 1989 and prior to the peak in case incidence. However, of these only 8,000 were estimated to have been in the last year of incubation, when it is hypothesized that affected tissue is at its most infectious. In the period $1990 - 95$ an estimated 299,000 $(285,000 - 317,000)$ infected animals were slaughtered for meat, with 43,500 being in the last year of the incubation period.

The basic reproduction number corresponding to feed-borne transmission, $R_0^{(FF)}$, is the expected number of secondary feed-borne infections generated by a single primary feed-borne infection occurring in an entirely susceptible population (Section 4.6). Assuming that feed-borne infectivity increases exponentially throughout the incubation period, the estimated annual mean value of $R_0^{(FF)}$ remains relatively consistent for Great Britain prior to the introduction of the ruminant feed ban in July 1988 (Figure 5.5). From 1989 onward, $R_0^{(FF)}$ was estimated to be below the unit threshold even though the average number of animals infected by a maximally infectious bovine in an entirely susceptible population, $I^{(FF)}$, remained above 5 for some years.

Assuming that feed-borne infectivity remains constant throughout the incubation period results in much lower estimates of $R_0^{(FF)}$ and $I^{(FF)}$ but, as in the case of exponentially rising infectivity, $R_0^{(FF)}$ was estimated to be below the unit threshold from 1989 onward (Figure 5.6). This is still the case even if non-reported case carcasses, in addition to infected animals slaughtered prior to the clinical onset of disease, were assumed to be recycled for cattle feed (Figure 5.6). These results indicate that if effectiveness of control measures is maintained, BSE will be unable to persist endemically in Great Britain.

5.4 Cross-validation of disease parameters

Although the per-capita incidence of reported BSE cases is an order of magnitude smaller in Northern Ireland compared with that observed in the rest of the United Kingdom, the overall patterns of the epidemics are similar. To validate the disease parameters (from the incubation period and age-dependent susceptibility/exposure distributions), the same model using parametric forms C7 was fitted to the Northern Ireland incidence data using disease parameters estimated from the incidence data Great Britain, but estimating the feed risk parameters for Northern Ireland independently.

The parameter estimates obtained from the analysis of the Great Britain data fitted the Northern Ireland incidence data well ($X^2 = 164$). Figures 5.7 and 5.8 give an indication of the quality of the fit of model. When the model with distributional forms C7 was fitted to the Northern Ireland incidence data without using previously estimated parameters ($X^2 = 90.9$), the mean incubation period was estimated to be shorter (4.73 years) than in Great Britain (5 years), resulting in lower estimated incidence of infections.

An estimated 12,300 animals were infected in Northern Ireland prior to 1997 (Ferguson et al., 1998). As in the rest of the United Kingdom, the infection incidence was estimated to peak earlier and be of a higher magnitude than the incidence of reported cases (Figure 5.9). Each 95% confidence interval presented is based on the minimum and maximum value obtained for that function of model parameters over the range of parameter values within the

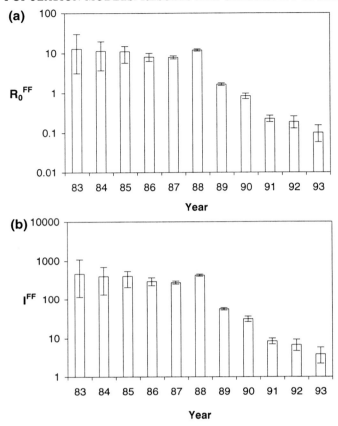

Figure 5.5 *Estimated annual averages of the basic reproduction number corresponding to feed-borne transmission, $R_0^{(FF)}$, and the average number of animals infected via feed by a maximally infectious bovine in an entirely susceptible population, $I^{(FF)}$, as a function of the time of slaughter of the primary infection assuming that feed-borne infectivity increases exponentially throughout the incubation period and that only pre-clinical animals were recycled for cattle feed (with 95% confidence intervals).*

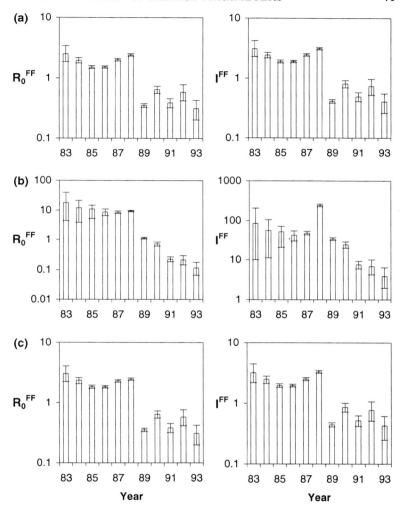

Figure 5.6 *Estimated annual averages of the basic reproduction number corresponding to feed-borne transmission, $R_0^{(FF)}$, and the average number of animals infected via feed by a maximally infectious bovine in an entirely susceptible population, $I^{(FF)}$, as a function of the time of slaughter of the primary infection assuming (a) that feed-borne infectivity remains constant throughout the incubation period with only pre-clinical animals being recycled for cattle feed; (b) exponentially rising and (c) constant infectivity with non-reported cases as well as pre-clinical animals being recycled for cattle feed (with 95% confidence intervals).*

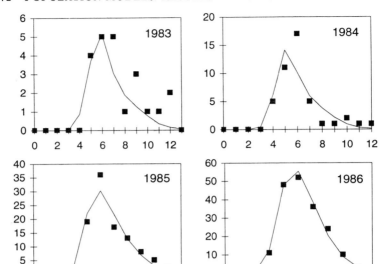

Figure 5.7 *The fitted incidence of reported BSE cases in Northern Ireland by birth cohort and age at onset for cohorts 1983 to 1986 with the observed case incidence (squares). The model assumed an incubation period distribution of form C and an age-dependent susceptibility/exposure distribution of form 7 and used the parameter estimates obtained from the analysis of the Great Britain incidence data.*

univariate 95% likelihood ratio confidence intervals, under the assumption that the risk of feed-borne infection drops to zero by July 1996.

5.5 Predictions

Predictions of future feed risk are critical to the prediction of future case and infection incidence. The past feed-risk profile is modelled using a spline with 20 knots, with geometric interpolation between adjacent knots. Feed risk through mid-1996 is extrapolated using simple linear regression from the estimated feed risk between mid-1991 and mid-1993, and beyond mid-1996 feed risk is assumed to

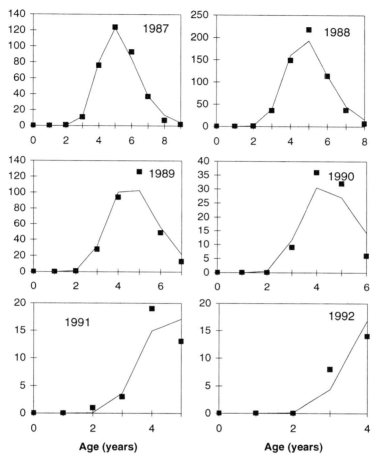

Figure 5.8 *The fitted incidence of reported BSE cases in Northern Ireland by birth cohort and age at onset for cohorts 1987 to 1992 with the observed case incidence (squares). The model assumed an incubation period distribution of form C and an age-dependent susceptibility/exposure distribution of form 7 and used the parameter estimates obtained from the analysis of the Great Britain incidence data.*

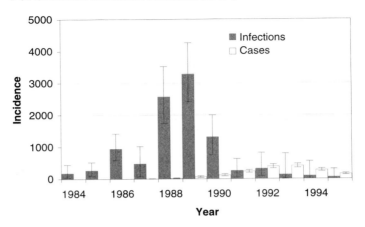

Figure 5.9 *Estimated annual incidence of infections and cases for North-ern Ireland assuming no maternal and no horizontal transmission (with 95% confidence intervals).*

be zero. The justification for this choice of date is that all mam-malian MBM was banned from use in livestock feed from 29 March 1996, and from 1 August 1996 it became an offence to possess feed containing mammalian MBM.

Although there is substantial variability in the predicted inci-dence between 1997 and 2001 over the 24 parametric combinations fitted to the data for Great Britain (Table 5.3), between 9,300 and 11,000 cases are predicted for the five models with likelihood ra-tio goodness-of-fit statistics less than 400 (assuming no maternal and no horizontal transmission). This variability, due largely to the uncertainty in the level of feed risk since 1992, is reduced by the inclusion of maternal transmission (Section 5.6), which helps explain the tail of the epidemic and thereby causes the estimated feed risk to drop to zero earlier.

Table 5.4 presents case and infection predictions (with 95% con-fidence intervals) over the years 1997 − 2001 for Great Britain and Northern Ireland, assuming no maternal or horizontal transmis-sion, for the model with an incubation period distribution of form C and an age-dependent susceptibility/exposure distribution of form 7. In all cases the models were constrained so that the risk of feed-borne infection drops to zero by July 1996.

Table 5.4 *The predicted incidence of new infections and cases for the years 1996 to 2001 in Great Britain (based on data through mid-1996) and Northern Ireland (based on data through mid-1997) for the model with no maternal and no horizontal transmission (with 95% confidence intervals).*

Year	Great Britain		Northern Ireland	
	Infections	Cases	Infections	Cases
1996	0	8452	14	74
	(0,533)	(7673,9355)	(0,115)	(48,122)
1997	0	5125	0	46
	(0,0)	(4136,6381)	(0,0)	(23,154)
1998	0	2628	0	27
	(0,0)	(1819,4011)	(0,0)	(6,158)
1999	0	1090	0	16
	(0,0)	(659,2169)	(0,0)	(1,115)
2000	0	380	0	7
	(0,0)	(211,943)	(0,0)	(0,58)
2001	0	118	0	2
	(0,0)	(63,337)	(0,0)	(0,19)

5.6 Maternal transmission

Assuming that maternal infectiousness is restricted to the late stages of the maternal incubation period (see Chapter 6), the incidence of maternal infections tracks the incidence of BSE cases in the cattle population. Thus, it is the tail of the case epidemic that is most affected by this transmission route, since maternal transmission only reaches substantial levels toward 1992, well after the peak of feed-borne infection incidence.

In the results presented here, the infectiousness function $\Omega_M(v)$ is assumed to be a step function, *i.e.* animals have a probability β_M of transmitting BSE to their unborn offspring if they are within time ω_M of the onset of clinical signs of disease. Within the back-calculation model, maternal transmission rates are calculated

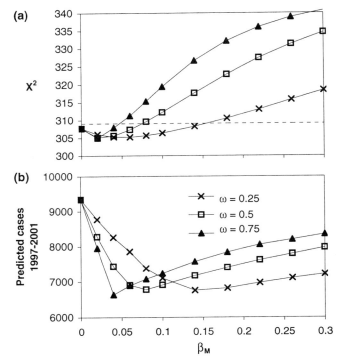

Figure 5.10 *The sensitivity to the probability of maternal transmission,* β_M, *of (a) the likelihood ratio goodness-of-fit statistic,* X^2, *(the dashed line indicates the 95% confidence level for 1 degree of freedom, relative to the minimum of the* X^2 *profile) and (b) the predicted incidence of cases in 1997 to 2001.*

iteratively (4.24) with results given accurately up to third order in β_M (Section 4.7).

Assuming an age-dependent susceptibility/exposure distribution of form 7 and an incubation period distribution of form C, the best fitting model assuming no maternal or horizontal transmission, from the analysis of incidence data in Great Britain, the 95% confidence interval for β_M is between 0 and 0.16 for $\omega_M = 0.25$ years, 0 and 0.08 for $\omega_M = 0.5$ years, and 0 and 0.04 for $\omega_M = 0.75$ years. Within these bounds for $\omega_M = 0.5$ years, the estimated incidence of maternal infections between 1974 and 1995 and the case incidence between 1997 and 2001 due to maternal transmission increases linearly from 0 to 4,200 and 0 to 460, respectively.

The best estimates of the rate and duration of maternal transmission, obtained from the analysis of incidence data in Great Britain, were $4 - 10\%$ and $3 - 6$ months, respectively (Figure 5.10), in line with the best estimates of maternal transmission from more detailed studies (see Chapter 7), though the inclusion of maternal transmission did not significantly improve the fit of the model to the data. The confidence interval for the rate of maternal transmission obtained from the Northern Ireland data, 1% to over 50%, was uninformative (Ferguson *et al.*, 1998).

Maternal transmission at such levels actually decreases both the estimated incidence of infections between 1974 and 1995 and predicted case incidence between 1997 and 2001 (Figure 5.10 and Table 5.5), since maternal transmission helps explain the tail of the epidemic and thereby causes the estimated feed risk to drop to zero earlier (Figure 5.10). Thus, the assumption of no maternal transmission can be used to obtain a robust upper bound on both past infection incidence and future case incidence.

5.7 Horizontal transmission

After the cessation of feed-borne transmission, the only infection route that could lead to endemic persistence of BSE is direct horizontal transmission. The level of horizontal transmission is determined by the transmission coefficient, β_H, the age-dependent susceptibility/exposure of an animal to direct horizontal transmission, $g_H(a)$ and the incubation-stage-dependent infectiousness of an infected animal, $\Omega_H(v)$. For simplicity, we assume that neither age-dependent susceptibility/exposure nor the incubation period depends on the route of transmission (feed-borne or direct horizontal). We assume the infectiousness function $\Omega_H(v)$ to be a step function, *i.e.* animals are infectious through direct horizontal transmission only if they are within time ω_H of the onset of clinical signs of disease.

The incidence of direct horizontally transmitted infections, like maternally transmitted infections, tracks the incidence of BSE cases in the cattle population if transmission is restricted to the late stages of the incubation period. This leads to the cases arising from non-feed-borne infections being concentrated later in the epidemic. On the other hand, if there is constant infectivity throughout the

Table 5.5 *The predicted incidence of new infections and cases for the years 1996 to 2001 in Great Britain (based on data through mid-1996) and Northern Ireland (based on data through mid-1997) for the model with 10% maternal transmission for the last 6 months of the maternal incubation period and no horizontal transmission (with 95% confidence intervals).*

	Great Britain		Northern Ireland	
Year	Infections	Cases	Infections	Cases
1996	212	8075	9	73
	(188,241)	(7356,8944)	(4,112)	(47,121)
1997	100	4197	1	40
	(84,119)	(3583,4944)	(1,6)	(20,148)
1998	39	1741	1	19
	(33,47)	(1450,2098)	(0,5)	(9,149)
1999	14	641	0	4
	(12,17)	(534,772)	(0,3)	(2,54)
2000	5	235	0	1
	(4,6)	(198,280)	(0,1)	(1,19)
2001	2	89	0	0
	(2,2)	(76,105)	(0,0)	(0,5)

incubation period then there is more synchrony between the cases arising from feed-borne and non-feed-borne infections.

The estimate of the basic reproduction number corresponding to horizontal transmission, $R_0^{(HH)}$, does not depend on time, if the birth rate is constant (Section 4.6). Making this simplifying assumption in the analysis of incidence data in Great Britain, the upper 95% confidence bound for $R_0^{(HH)}$ is approximately 0.16 for $\omega_H = 0.5$ (Figure 5.11) and 0.09 for $\omega_H \to \infty$, far below the unit threshold for disease persistence.

The upper 95% confidence bound for $R_0^{(HH)}$, 0.16 (for $\omega_H = 0.5$ years), corresponds to approximately 141,000 horizontally transmitted infections among the estimated 979,000 infections between between 1974 and 1995 (Figure 5.11) in Great Britain, but like

Table 5.6 *The predicted incidence of new infections and cases for the years 1996 to 2001 in Great Britain (based on data through mid-1996) and Northern Ireland (based on data through mid-1997) for the model with horizontal transmission (with $\beta_H = 6.0$) for the last 6 months of the incubation period and no maternal transmission (with 95% confidence intervals).*

| | Great Britain | | Northern Ireland | |
Year	Infections	Cases	Infections	Cases
1996	6088	8654	55	76
	(5918,6721)	(8383,9440)	(44,120)	(65,123)
1997	4019	5792	42	55
	(3912,4415)	(5624,6413)	(33,84)	(48,112)
1998	2581	3765	28	41
	(2514,2770)	(3683,4069)	(22,55)	(35,80)
1999	1644	2409	16	25
	(1599,1756)	(2356,2569)	(12,32)	(21,48)
2000	1050	1555	9	14
	(1019,1133)	(1517,1669)	(7,19)	(12,29)
2001	661	1012	6	8
	(640, 718)	(988,1094)	(4,12)	(7,18)

animals infected through other transmission routes, most horizontally infected animals do not survive to become cases because of the form of the survival distribution. However, unlike maternal transmission, the presence of horizontal transmission increases the predicted incidence of cases between 1997 and 2001 in Great Britain from 9,300 (assuming no horizontal transmission) to 15,300 for $R_0^{(HH)}$=0.16 and $\omega_H = 0.5$, corresponding to 86% of the predicted case incidence between 1997 and 2001 (Figure 5.11 and Table 5.6). If infectivity is assumed to be constant throughout the incubation period, $\omega_H \to \infty$, then the predicted incidence of cases between 1997 and 2001 in Great Britain is bounded by 10,700, corresponding to $R_0^{(HH)}$=0.09.

As anticipated, given the rapid decline in BSE case incidence in

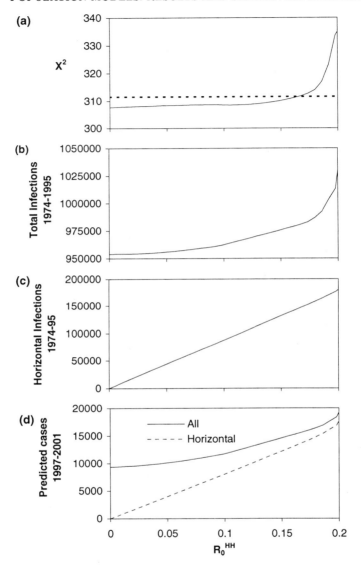

Figure 5.11 *The sensitivity to $R_0^{(HH)}$ of (a) the likelihood ratio goodness-of-fit statistic, X^2, (the dashed line indicates the 95% confidence level for 1 degree of freedom, relative to the minimum of the X^2 profile), (b) the estimated incidence of infections due to horizontal transmission in the years 1974 to 1995, (c) the estimated total incidence of infections in the years 1974 to 1995 and (d) the predicted incidence of cases to arise in 1997 to 2001 in Great Britain.*

recent years, the population-level data provide no evidence that direct horizontal transmission could sustain the epidemic. Even under the extreme condition that only 20% of the national herd was ever exposed/susceptible to BSE, the upper 95% confidence bound for $R_0^{(HH)}$, 0.75, is still well below the persistence threshold. However, population level analyses alone can never exclude the possibility that a small core group of animals exist for which $R_0^{(HH)} > 1$.

5.8 Under-reporting

One of the most difficult aspects of modelling the BSE epidemic is estimating the rate of under-reporting, due to the complex interdependency of the estimates of under-reporting, the incubation period, and age-dependent susceptibility/exposure distribution. In this section, we characterize the effects of under-reporting profiles on model fit and address the confounding factors that complicate any simple interpretation of the results.

Since no independent data on reporting rates are available, it is impossible to fit a time-dependent probability of reporting, $\Lambda(t)$, across the whole epidemic. It is therefore necessary to assume a date at which, or beyond which, reporting is complete. Unless otherwise indicated, we assume that cases were fully reported after the introduction of compulsory notification of BSE cases in mid-1988.

Under-reporting was modelled to allow very high under-reporting in the early years of the epidemic when BSE was largely unknown and had not been diagnosed by most veterinarians. The parametric form

$$\Lambda(t) = \begin{cases} \frac{1}{1+1.5\delta_1+[\delta_2(1987-t)]^{1.2}}, & t < t_1 \\ \frac{1}{1+\delta_1(1988.5-t)}, & t_1 \leq t < 1988.5 \\ 1, & t \geq 1988.5 \end{cases} \qquad (5.1)$$

was fitted to the data with $t_1 = 1987$ and in a simplified form (without parameter δ_2) with $t_1 = 1974$. In the analysis of the incidence data from Great Britain, the likelihood ratio goodness-of-fit statistic for model C7 with the simplified form of under-reporting was greater than 400 for all values of δ_1. Assuming $t_1 = 1987$, a dramatic fall in the likelihood ratio goodness-of-fit statistic is seen, as δ_2 increases to 100 (Ferguson et al., 1997a). The maximum likelihood value of δ_2 is 244 but for all values of $\delta_2 > 100$ estimates of the incidence of total infections 1974−1995, the mean incubation period, and the mean of the age-dependent susceptibility/exposure distribution as well the predicted incidence of cases between 1997 and 2001 are very similar. Under-reporting was modelled using (5.1) with $t_1 = 1987$ for other model fits discussed.

The under-reporting model allows the fitted total case incidence profiles for the cohorts 1981 − 1986 (Figure 5.12) to show a relatively constant age structure of cases despite the dramatic differences in the fitted (and observed) reported case incidence profiles. For such dramatic changes to the age-structure of the epidemic to arise from the increasing force of infection in the early years of the epidemic, a much flatter age-dependent susceptibility/exposure distribution would be required. An alternative hypothesis is that the incubation period of BSE dramatically shortened in the initial stages of the epidemic, due to changes in the mean infective dose of the BSE aetiological agent or to the passaging of infection through cattle, and then stabilized.

The inclusion of an additional parameter corresponding to under-reporting in the period July 1988 to February 1990 (when 100% compensation for BSE cases was introduced) had a negligible impact on the goodness-of-fit to the resampled incidence data from Great Britain, but produced a significant improvement in the fit of the model to the raw incidence data with an estimated 13% of cases being unreported during this time. This discrepancy arises due to the similar effects of resampling (see Section 3.2.1) and under-reporting of slowing the changes in the age-at-onset distribution from cohort to cohort. In the case of resampling, this comes about because a proportion of animals is reassigned to the previous cohort, thereby boosting the numbers of cases seen at relatively young ages at onset (4 − 6 years) across the 1981 − 85 cohorts (since epidemic growth is rapid across those cohorts).

5.9 Conclusions

The estimated incidence of past infections and the predicted incidence of cases have been shown to be robust to changes in nearly all model parameters and distributions. Estimates of the past number and pattern of BSE infections prove relatively insensitive to variation in model parameters. All the models presented in this work produce estimates of total infections in Great Britain in the range 920,000−1,050,000.

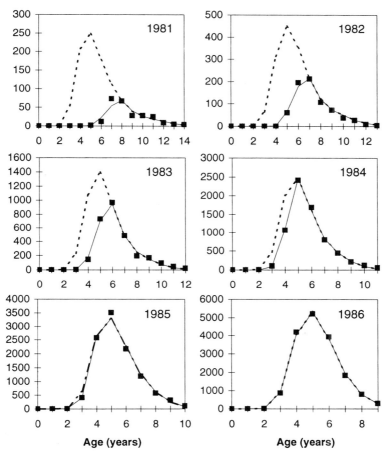

Figure 5.12 *The fitted incidence of reported (solid lines) and total (dotted lines) BSE cases in Great Britain by birth cohort and age at onset for cohorts 1981 to 1986 with the observed case incidence (squares). The model assumed an incubation period distribution of form C and an age-dependent susceptibility/exposure distribution of form 7.*

Clearly, prediction is more difficult due to the intrinsic uncertainties involved in extrapolating the estimated feed risk profile. However, all models with likelihood ratio goodness-of-fit statistics less than or equal to 400 for fits to the data in Great Britain available in 1997 $(\min(X^2) = 308$ corresponding to model C7) predict no more than 11,000 cases of BSE between 1997 and 2001. Furthermore, these predictions can be regarded as conservatively high since the inclusion of low levels of maternal transmission decreases predictions of future case incidence,

We cannot exclude the possibility of direct horizontal transmission, but on the basis of analyses of its possible contribution to the overall basic reproduction number, R_0, of the agent throughout the course of the epidemic, we believe that even if such transmission does occur, its magnitude is insufficient to maintain BSE endemically in the British cattle herd (Ferguson et al., 1999b; Woolhouse and Anderson, 1997). Ideally, carefully designed and controlled experiments are needed to assess the likelihood of horizontal transmission, but these may not be possible at this late stage of the epidemic. Furthermore, even if set up in the near future, results would not be available for analysis for at least five to six years due to the long incubation period of the disease and the absence of an ante-mortem test for the presence of the aetiological agent.

Insight into the past pattern of human exposure to BSE-infected material can be gained from the numbers of infected animals slaughtered through time stratified by their incubation stage. The relevance of this stratification is that it is thought that tissue from affected animals is at its most infectious around the time of disease onset. The estimated pattern reflects the numbers quoted above: while very large numbers of infected animals were slaughtered, most were in the early stages of incubation, especially before November 1989 when the ban on specified offal from cattle entering human food was introduced. As more information on disease pathogenesis in cattle becomes available, estimates of the number of infected cattle, stratified by incubation stage, which were slaughtered for consumption, will enable refinement in the estimates of past human (relative) exposure to be made. Given a longer time-series of the incidence of new variant CJD in humans, this may facilitate more reliable prediction of the overall scale of any future human epidemic.

CHAPTER 6

Individual survival models

6.1 Introduction

Chapter 4 developed models describing the age and temporal structure of an epidemic process at the level of an entire population.
While we modelled the probability that a single host would develop disease by a certain age, at any given time all hosts in the
population were assumed to experience the same infection hazard,
and the model was fitted to aggregated case data. The advantage
of such models is their relative parsimony in the number of parameters fitted; the disadvantage is that they are unable to capture
the heterogeneity in exposure and the resulting variation in case
incidence at the regional or farm level.

A variety of approaches can be adopted to model epidemiological heterogeneity at the sub-population (e.g. region, herd or individual) level, depending on the questions being addressed. To explore the mechanisms giving rise to disease clustering (whether in
space or in distinct sub-populations), stochastic models that capture the essential elements of the feed-borne transmission route are
required. Such models are discussed in Chapters 8 and 9. Alternatively, if one wishes to explore the effect of individual-specific risk
factors, models describing individual disease risk and survival are
required. It is this type of model that is developed in this chapter.

A model of individual risk must by definition incorporate information on the overall infection hazard experienced by hosts within
the same sub-population: ideally, therefore, such a model would
incorporate key elements of a mechanistic model of transmission
dynamics. In practice, however, this is impractical due to limitations in computational resources and the difficulties inherent in
fitting such stochastic non-linear models to time-series data. It is
therefore necessary to estimate the infection hazard experienced
by a sub-population from available disease incidence data.

The main application of this type of modelling to BSE epidemiology has been in the analysis of the nature and magnitude of a

maternally enhanced risk of disease. Maternal risk enhancement might be of two forms: direct infection of a calf by its dam or inheritance of a genetically determined higher level of susceptibility to infection. We apply the models developed to two different data sets. The first is a seven-year cohort study (Wilesmith et al., 1997; Donnelly et al., 1997c; Gore et al., 1997; Curnow et al., 1997) of the offspring of BSE-affected dams and their age- and herd-matched controls, for which complete survival data are available. The second is identification data on the dams of BSE cases that allows the pairs of dams and calves both of which developed BSE to be linked using the BSE database (Donnelly et al., 1997a). Complete survival data on animals that did not develop BSE is not available in this case, however.

To go beyond descriptive statistical summaries of the data and obtain estimates of key parameters, it is necessary to develop mechanistic survival models that model the transmission process. The modelling approach adopted incorporates simultaneously (i) the rate of maternal transmission and (ii) measures of genetically variable susceptibility to feed-borne infection, and represents a generalization of the models presented by Donnelly et al. (1997c) and Ferguson et al. (1997b).

6.2 Maternal risk enhancement models

Case data on the disease status of dams and offspring fall into four classes: both affected (11); both unaffected (00); only dam affected (10); only offspring affected (01). We here construct a likelihood model representing the probability of these four states that incorporates age at clinical onset of disease, censoring, and subpopulation (herd) infection hazards.

The unconditional likelihood for the (11) state takes the form of a joint density on the ages at disease onset of the dam and offspring, u' and u respectively: i.e. $L_{11}(u', u)$. For other states, one or both individuals do not develop disease, in which case c or c' are taken to represent the age of censoring of the dam and offspring, respectively, (which may or may not be due to slaughter) The likelihoods $L_{01}(c', u)$ and $L_{10}(u', c)$ are therefore densities in u or u' and cumulative densities in c' or c, while $L_{00}(c', c)$ is a cumulative density in both c and c'. These likelihoods are related thus

$$L_{00}(c',c) = L_{\cdot 0}(c) - \int_0^{c'} L_{10}(u',c)du'$$

$$= L_{0\cdot}(c') - \int_0^c L_{01}(c',u)du,$$

$$L_{10}(u',c) = L_{1\cdot}(u') - \int_0^c L_{11}(u',u)du,$$

$$L_{01}(c',u) = L_{\cdot 1}(u) - \int_0^{c'} L_{11}(u',u)du',$$

where $L_{1\cdot}(u')$ and $L_{\cdot 1}(u)$ are the univariate densities for the ages of onset of the dam and offspring respectively, and $L_{0\cdot}(c')$ and $L_{\cdot 0}(c)$ are the corresponding cumulative densities of no disease onset by ages c' and c. These are themselves related by

$$L_{\cdot 0}(c) = 1 - \int_0^c L_{\cdot 1}(u)du$$

$$L_{0\cdot}(c') = 1 - \int_0^{c'} L_{1\cdot}(u')du'.$$

The univariate densities are required here since we are using unconditional joint likelihoods, so that, for example,

$$L_{10}(u',c) \neq \int_c^\infty L_{11}(u',u)du,$$

since a host not experiencing disease onset by a given age may never have been infected.

We therefore only need to formulate three likelihood densities: $L_{11}(u',u)$, $L_{1\cdot}(u')$ and $L_{\cdot 1}(u)$. Note that all these densities are dependent on the times of birth of both dam and offspring, t_0' and t_0. The univariate densities are just given by a herd-specific version of equation (4.25), $\phi_C^h(t_0,u)$ for herd h. Thus

$$L_{\cdot 1}(u) = \phi_C^h(t_0,u)$$

$$L_{1\cdot}(u') = \frac{\phi_C^h(t_0',u')}{1 - \int_0^{t_0-t_0'} \phi_C^h(t_0',a)da}.$$

Note that $L_{1\cdot}(u')$ is defined to be conditional on the animal's surviving to produce an offspring at time t_0. It is assumed here that dams will not give birth after disease onset; this is a slight approximation for BSE since calves have been born to dams within a few weeks following disease onset.

The maternal transmission rate contributes to the likelihood for dam−calf pairs in which the dam was infected with the aetiological agent of BSE before the birth of the calf. Since age of infection data are never available for naturally infected animals, the likelihood term for a dam−calf pair in which both animals experience disease onset, $L_{11}(u', u)$, must incorporate the possibilities that the dam was infected before the birth and thus capable, at least theoretically, of maternally transmitting the infectious agent and that the dam was infected after the birth. Similarly, a dam not observed to experience the onset of clinical signs of BSE cannot be assumed to have remained uninfected with the aetiological agent of the disease.

We therefore need to refine our formulation of the probability density of the age of disease onset, $\phi_C(t_0, u)$ (4.25) to allow restriction of the possible range of infection ages, (a_1, a_2), and to stratify by susceptibility class and herd, thus:

$$\phi_C^{hg}(t_0, u | a_1, a_2) = S(u) p_g \int_{a_1}^{a_2} \sum_{j=F,M,H} \rho_{jg}^h(u - x | t_0) f_j(x) dx \quad (6.1)$$

where g is the susceptibility class of the host. We make the simplifying assumption throughout that hosts are in the same holding throughout their lives.

When describing the infection hazard experienced by the offspring of a specific animal, we need to consider the maternal transmission route separately from other routes in (6.1) since the identity (and later disease status) of the dam is known. The probability of maternal transmission from an infected host with disease onset at time v after the birth is $\beta_M \Omega_M(v)$, while the probability density for the age at disease onset of a host of susceptibility class g infected via feed or horizontal transmission is

$$\Phi^{hg}(t_0, u, a_r) = S(u) \int_{u-a_r}^{u} \sum_{j=F,H} \rho_{jg}^h(u - x | t_0) f_j(x) dx$$

where a_r is the age beyond which no infection hazard is experienced, set to u if the individual is exposed until onset.

The likelihood $L_{11}(u', u)$ is then given by

$$L_{11}(u',u) = \sum_{g,g'} \frac{p_{Mg|g'}}{1 - \int_0^{t_0-t_0'} \phi_C^{hg'}(t_0',a)da} \left(\phi_C^{hg'}(t_0',u'|0,t_0-t_0') \right.$$
$$\times \{\beta_M \Omega_M(t_0'+u'-t_0)f_M(u)$$
$$+ [1 - \beta_M \Omega_M(t_0'+u'-t_0)]\Phi^{hg}(t_0,u,a_r)\}$$
$$\left. + \phi_C^{hg'}(t_0',u'|t_0-t_0',u')\Phi^{hg}(t_0,u,a_r) \right).$$

Hence

$$L_{11}(u',u) = \sum_{g,g'} \frac{p_{Mg|g'}}{1 - \int_0^{t_0-t_0'} \phi_C^{hg'}(t_0',a)da}$$
$$\times \left\{ \phi_C^{hg'}(t_0',u'|0,u')\Phi^{hg}(t_0,u,a_r) \right.$$
$$+ \phi_C^{hg'}(t_0',u'|0,t_0-t_0')\beta_M \Omega_M(t_0'+u'-t_0)$$
$$\left. \times [f_M(u) - \Phi^{hg}(t_0,u,a_r)] \right\}.$$

We make use of simplified versions of these conditional likelihood terms in the analysis of data from the maternal cohort study and dam identification for BSE cases born after the introduction of the ruminant feed ban. In particular, we ignore the possibility that a dam was infected after the birth of the calf but prior to age u' (effectively assuming $\int_{u'}^{\infty} \xi_{g'h}(u''|u'',t_0-t_0',t_0')du'' = 0$). These terms are negligible using the maximum likelihood estimate of $g(a)$ due to the much reduced susceptibility to feed-borne infection after 2 years of age, the typical age at first lactation.

Although population genetics models can be used in a straight-forward manner to obtain the inheritance probabilities, $p_{Mg,g'}$, the extent to which the distribution of genotypes of BSE-affected (and the genotypes of unaffected) dams differs from the Hardy–Weinberg equilibrium distribution of genotypes depends upon the infection hazard experienced by those dams. Thus, the dam geno-type probabilities, $p_{g'}$, will vary by herd if exposure varied by herd.

6.3 Analysis of the maternal cohort study

For ease of reference we reproduce again the basic results of the maternal cohort study (Table 6.1) in terms of the disease status outcomes of the animal pairs.

Since some animals were diagnosed with the clinical signs of BSE, while others did not show signs of obvious disease but were

Table 6.1 *Contingency table of the pair outcomes from the maternal cohort study in terms of histopathological examination of brain tissue and the onset of clinical signs of disease.*

Control		Maternally exposed		
		Hist. pos. Signs 1	Hist. pos. No signs 0_+	Hist. neg. No signs 0_-
Hist pos./Signs	1	6	0	6
Hist. pos./No signs	0_+	0	0	1
Hist. neg./No signs	0_-	27	9	252

only diagnosed after post-mortem histopathological examination at the end of the study, the likelihood model adopted below needs to distinguish between two types of asymptomatic outcomes in calves: histopathological positives (and therefore certainly infected) - 0_+, and histopathological negatives (and therefore probably uninfected) - 0_-. Using the notation adopted above, we define likelihood densities L_{00_-}, L_{00_+}, L_{10_-} and L_{10_+} thus:

$$L_{00}(c', c) = L_{00_-}(c', c) + L_{00_+}(c', c)$$
$$L_{10}(u', c) = L_{10_-}(u', c) + L_{10_+}(u', c)$$
$$L_{00_-}(c', c) = \int_u^\infty L_{01}(c', u) du$$
$$L_{10_-}(u', c) = \int_u^\infty L_{11}(u', u) du$$

Given the design of the study, the likelihoods of the different pair outcomes need to be defined conditionally on the disease status of the dam. These are shown in Table 6.2.

From the multinomial pair likelihood function, maximum likelihood estimates can be obtained for the parameters using direction-set techniques. The simultaneous likelihood ratio confidence region for all of the maximum likelihood estimated parameters contains all combinations of parameters that provide a similar goodness-of-fit to the observed data, as measured by the likelihood ratio statistic.

Table 6.2 *Likelihood expressions for pair outcomes from the maternal cohort study.*

	Control	Maternally exposed Hist. pos. Signs 1	Hist. pos. No signs 0_+	Hist. neg. No signs 0_-
Hist. pos./Signs	1	$\frac{L_{11}}{L_{1\cdot}}\frac{L_{01}}{L_{0\cdot}}$	$\frac{L_{10+}}{L_{1\cdot}}\frac{L_{01}}{L_{0\cdot}}$	$\frac{L_{10-}}{L_{1\cdot}}\frac{L_{01}}{L_{0\cdot}}$
Hist. pos./No signs	0_+	$\frac{L_{11}}{L_{1\cdot}}\frac{L_{00+}}{L_{0\cdot}}$	$\frac{L_{10+}}{L_{1\cdot}}\frac{L_{00+}}{L_{0\cdot}}$	$\frac{L_{10-}}{L_{1\cdot}}\frac{L_{00+}}{L_{0\cdot}}$
Hist. neg./No signs	0_-	$\frac{L_{11}}{L_{1\cdot}}\frac{L_{00-}}{L_{0\cdot}}$	$\frac{L_{10+}}{L_{1\cdot}}\frac{L_{00-}}{L_{0\cdot}}$	$\frac{L_{10-}}{L_{1\cdot}}\frac{L_{00-}}{L_{0\cdot}}$

6.4 Dam–calf pairs in the GB case database

Tracing BSE cases born following the introduction of the ban on the use of ruminant material in cattle feed in July 1988 (BSE Order, 1988) has identified 1346 BSE-affected dam–calf pairs. The identification data on the dams of BSE cases born after the introduction of the ruminant feed ban in July 1988 ('Born After Ban' cases or BABs) yields further information about the extent and pattern of maternal risk enhancement. While these data provide information on calves born earlier in the dam incubation periods than can be obtained from the maternal cohort study, the data are observational. For this reason, additional uncertainties including survival probabilities arise. As in the maternal cohort study, the continued use of feed contaminated with the aetiological agent of BSE after the introduction of the feed ban in July 1988 necessitates the modelling of feed-borne as well as maternal infections.

Let D_{hik}^{+} be the number of dams in holding h giving birth to animals in birth cohort i in BSE incubation stage k and D_{hi} be the total number of dams in holding h giving birth to animals in birth cohort i. Similarly, let B_{hi} denote the number of BABs in holding h and birth cohort i and B_{hi}^{I} denote the number of such animals with identified dams. Finally, let O_{hik} denote the number of BABs with identified BSE-confirmed dams in holding h born in birth cohort i with dams in BSE incubation stage k. The value of O_{hik} is bounded above by the minimum of B_{hi}^{I} and D_{hik}^{+} since the number of identified dam−calf positive pairs cannot be greater than the number of BABs or the number of BSE-confirmed dams in incubation stage k.

Assuming that, within a holding, BSE incidence in the calves is independent of that in the dams, the expected number of identified dam positive−calf positive pairs (P_k) for dam incubation stage k can be estimated by the number of BABs with identified dams weighted by the proportion of dams confirmed as BSE cases:

$$\widehat{P}_k = \sum_{h,i} \frac{D^+_{hik}}{D_{hi}} B^I_{hi} \tag{6.2}$$

In this manner, the indirect association between BSE in the dam and the calf caused by the clustering of BSE cases in holdings (Donnelly et al., 1997b) is accounted for by calculating the expected value within each holding and then summing over all holdings. Alternatively, the expected number of identified dam positive−calf positive pairs for dam incubation stage k can be estimated as:

$$\widetilde{P}_k = \frac{\sum_{h,i} B^I_{hi}}{\sum_{h,i} B_{hi}} \sum_{h,i} \frac{D^+_{hik}}{D_{hi}} B_{hi}. \tag{6.3}$$

The estimates obtained from (6.2) and (6.3) will be similar if the probability that a dam was identified is independent of the disease prevalence in the dams. Significant differences between these two methods would indicate biased identification rates of dams.

The corresponding observed to expected ratios of identified dam positive−calf positive pairs,

$$\widehat{R}_k = \frac{\sum_{h,i} O_{hik}}{\widehat{P}_k} \text{ and } \widetilde{R}_k = \frac{\sum_{h,i} O_{hik}}{\widetilde{P}_k},$$

can be compared with the mean, denoted $\bar{r}_k$, of the highly skewed distribution of within holding and calf birth cohort ratios of observed to expected numbers of positive pairs:

$$r_{hik} = \frac{O_{hik}}{D^+_{hik} B^I_{hi} / D_{hi}}.$$

Ratios significantly greater than 1 indicate a significant enhanced risk of BSE occurring in calves born to dams who subsequently developed BSE.

A bootstrap algorithm was used to generate confidence intervals for $\widehat{R}_k$ and $\widetilde{R}_k$ (DiCiccio and Ephron, 1996) since analytical characterization of the distributions generating these estimators proved intractable. The same method was used for $\bar{r}_k$ to avoid reliance on the assumption that the sample means of the r_{hik} values are normally distributed. Holdings were sampled with replacement from the database.

To go beyond these descriptive statistics and estimate the parameters of interest, it is necessary to fit mechanistic survival models. However, given the quantity of data, fitting a full likelihood survival model to individual animals would be highly computationally intensive. Instead we consider a simplified approach that fits the data aggregated at the holding level to retain the variable feed-risk (thereby controlling for clustering of dam–calf positive pairs by holding), but uses a simplified survival distribution.

We model the probability that the identified dam of the rth BAB was at incubation stage k (where $k = 0, \ldots, K$) at the time of calving. This probability depends on the number of eligible dams at incubation stage k at calving and for an animal in holding h and birth cohort i, $D^+_{hik}(r)$, for $k = 0, \ldots, K$ as well as the number of dams not in incubation stages $0, \ldots, K$, $D^-_{hi}(r)$, where

$$D^-_{hi}(r) = \text{int}[\gamma D_{hi}(r)] - \sum_{k=0}^{K} D^+_{hik}(r),$$

$1/\gamma$ is the factor by which the holding sizes are underestimated, and the function $\text{int}(z)$ rounds z to the nearest integer. We ignore ages at disease onset for both dams and calves (except as the age at disease onset in the dam determines the incubation stage of the dam at the time of calving) and assume that all infected calves have a probability of survival till disease onset which is determined only by the holding, birth cohort and whether the dam is affected by BSE. The probability that the identified dam of a BAB was at incubation stage k at the time of calving is given by

$$\zeta_{hi}\left(k | D^-_{hi}(r), D^+_{hik}(r) \; \forall k\right) =$$

$$D^+_{hik}(r) \int_0^{a_C} \int_{t'_0 + u' - t_0 \in k} \frac{L_{11}(u', u)}{L_{1.}(u')} du' \, du$$

$$\Bigg/ \Bigg(D^-_{hi}(r) \int_0^{a_C} \frac{L_{01}(a_C, u)}{L_{0.}(a_C)} du$$

$$+ \sum_{k=0}^{K} D^+_{hik}(r) \int_0^{a_C} \int_{t'_0 + u' - t_0 \in k} \frac{L_{11}(u', u)}{L_{1.}(u')} du' \, du \Bigg)$$

for $k = 0, \ldots, K$. For ease of notation in defining the likelihood, we let $k = -1$ denote that the dam was not recorded to experience the onset of BSE. The probability that the identified dam of a BAB was not recorded to experience the onset of BSE is thus given by

$$\zeta_{hi} \left(k = -1 | D_{hi}^-(r), D_{hik}^+(r) \ \forall k\right) =$$

$$1 - \sum_{k=0}^{K} \zeta_{hi} \left(k | D_{hi}^-(r), D_{hik}^+(r) \ \forall k\right). \qquad (6.4)$$

For $r = 1$, the stratified numbers of eligible dams,

$$D_{hik}^+(r) = D_{hik}^+$$

whereas for $r > 1$ the numbers are conditional so that $D_{hik}^+(r) = D_{hik}^+ - \sum_{q=1}^{r-1} I_k(x_q)$ where $I_k(x_q) = 1$ when $x_q = k$ and 0 otherwise. Similarly, $D_{hi}(1) = D_{hi}$ and $D_{hi}(r) = D_{hi} - \sum_{k=0}^{K} \sum_{q=1}^{r-1} I_k(x_q)$ for $r > 1$.

Let m index the BABs with identified dams within a holding birth cohort such that $m = 1, \ldots, B_{hi}^I$ for holding h and birth cohort i, and let x_m equal the dam incubation stage at calving for the dam of calf m. The joint likelihood for all BABs with identified dams in holding h and birth cohort i (L_{hi}) is given by the sum over all possible orderings $(\mathbf{X})$ of the product of these conditional probabilities such that

$$L_{hi} = \sum_{x \in \mathbf{X}} \prod_{m=1}^{B_{hi}^I} \zeta_{hi}(x_m | x_1, \ldots, x_{m-1}, D_{hi}^-, D_{hik}^+ \ \forall k). \qquad (6.5)$$

The full likelihood is thus the product of the L_{hi} terms for all holdings h and all birth cohorts i.

Clearly it would be infeasible to estimate a holding- and birth-cohort-specific survival probability for each dam incubation stage. However, assuming that the survival probability depends only on whether the dam was observed to experience clinical onset, the model likelihood (6.5) depends only on the ratio, λ, of the probability of survival of a maternally exposed animal compared with that of the offspring of an unaffected dam.

We further simplify the model probability to be

$$\zeta_{hi} \left(k | D_{hi}^-(r), D_{hik}^+(r) \ \forall k\right) = \frac{\lambda D_{hik}^+(r) p_{hik}^+}{D_{hi}^-(r) p_{hi}^- + \lambda \sum_{k=0}^{K} D_{hik}^+(r) p_{hik}^+}$$

for $k = 0, \ldots, K$ where p_{hik}^+ is the probability that an animal born in herd h in birth cohort i to a BSE-affected dam in incubation stage k is infected and p_{hi}^- is the probability that an animal born to an unaffected dam is infected. Assuming a simple susceptibility class structure with all unaffected dams and their offspring in susceptibility class 1 with $s_1 = 1$ (Section 4.8) and all BSE-affected dams and their offspring in susceptibility class 2, we parametrize these probabilities as

$$p_{hik}^+ = \beta_{Mk} + (1 - \beta_{Mk})\left[1 - \exp(-r_{Fhi}s_2/\gamma)\right]$$

and

$$p_{hi}^- = 1 - \exp(-r_{Fhi}/\gamma)$$

where r_{Fhi} is the feed-borne infection hazard and, as previously defined, $1/\gamma$ is the factor by which the holding sizes are underestimated. The infection hazard is estimated as

$$\hat{r}_{Fhi} = \alpha \frac{\sum_i B_{hi}}{\sum_i D_{hi}} \frac{\sum_h B_{hi}}{\sum_{h,i} B_{hi}}$$

where α is the ratio of infections to cases.

Maximum likelihood estimates can be obtained for the susceptibility and maternal transmission parameters using direction-set techniques. The simultaneous likelihood ratio confidence region contains all combinations of parameters that provide a similar goodness-of-fit to the observed data, as measured by the likelihood ratio statistic. The goodness-of-fit of the model is measured by the comparison of the difference between the maximized model likelihood and the saturated data likelihood with the distribution of the analogous differences obtained from bootstrap samples from the model. The saturated likelihood is obtained from (6.4) and (6.5) where

$$\zeta_{hi}\left(k \mid D_{hi}^-(r), D_{hik}^+(r) \; \forall k\right) =$$

$$\frac{\lambda D_{hik}^+(r)\frac{O_{hik}(r)}{B_{hi}^I(r)}}{D_{hi}^-(r)\left(1 - \frac{O_{hik}(r)}{B_{hi}^I(r)}\right) + \lambda \sum_{k=0}^K D_{hik}^+(r)\frac{O_{hik}(r)}{B_{hi}^I(r)}} \tag{6.6}$$

for $k = 0, \ldots, K$. For $r = 1$, $O_{hik}(r) = O_{hik}$ and $B_{hi}^I(r) = B_{hi}^I$ whereas for $r > 1$ the numbers are conditional so that

$$O_{hik}(r) = O_{hik} - \sum_{q=1}^{r-1} I_k(x_q)$$

and

$$B_{hi}^{I}(r) = B_{hi}^{I} - \sum_{k=0}^{K} \sum_{q=1}^{r-1} I_k(x_q).$$

The parameters of key interest are the rate of maternal transmission by incubation stage and the genetic susceptibility ratios, s_g. To explore the model predictions, assuming the survival probabilities are identical for the offspring of BSE-affected and unaffected dams, we calculate the ratio of the expected number of dam−calf BSE pairs from the model to the expected number under the assumption of independence between cases of disease in the dams and calves. The former is estimated as the mean of bootstrap samples using the maximum likelihood parameter estimates of s_g and β_M and assuming $\lambda = 1$ (regardless of the value of λ assumed in the estimation). In the case of no association between dam BSE-status and calf BSE-status we would expect these ratios to take values close to 1, while if an association is present we expect these ratios to have values greater than 1.

6.5 Conclusions

In this chapter we have extended the population-level survival models to provide tools for the analysis of enhanced risk of BSE in the offspring of clinically affected dams. Although the first step is clearly to test for the significance of such enhancement, the interpretation of the results based only upon relative risks (or other descriptive measures) would be very limited. Models of transmission dynamics within the survival analysis framework allow biologically relevant parameters (maternal transmission rates and relative susceptibilities) to be estimated. In the next chapter we present applications of the models discussed here to data from the maternal cohort study and identification data on the dams of BSE cases.

CHAPTER 7

Maternal risk enhancement models: results

7.1 Introduction

The simplest and most powerful method to detect maternal transmission would be to house calving cattle exhibiting clinical signs of
BSE in previously unoccupied accommodation and to recruit their
calves just after calving, ensuring that these 'maternally exposed'
animals are never exposed to protein supplements containing potentially infectious MBM. Such a design would eliminate any risk
of feed-borne and horizontal infection, and any effect of genetically
variable susceptibility to these routes of infection. In studies of animals that have been potentially exposed to all routes of infection,
a key indicator of the existence of maternal transmission of BSE
capable of distinguishing it from genetically variable susceptibility
would be the existence of a trend in the magnitude of the maternally enhanced risk with the timing of the birth as a function of
the duration of infection, or incubation stage, of the dam. It is just
such a trend that we are able to detect in the data on dam−calf
pairs in the main epidemiological database and, to a more limited
extent, from the maternal cohort study.

Using the models developed in Chapter 6, in this chapter we
estimate the relative contributions to the observed maternally enhanced risk of BSE in the offspring of affected dams of direct maternal transmission of the aetiological agent of BSE and of genetically
variable susceptibility controlling for the substantial between-herd
variation in the risk of exposure to the BSE agent.

7.2 Maternal cohort study

To begin with, we consider models of genetically determined susceptibility to BSE in the absence of direct maternal transmission.
Such models, allowing for resistant animals and for single-locus

Table 7.1 *Genotype frequencies observed by Neibergs* et al. *(1994) in confirmed BSE cases (n = 56) and cattle unaffected by BSE (n = 38).*

Genotypes	Confirmed BSE	Unaffected cattle
A A	0.48	0.29
AB	0.16	0.05
AC	0.30	0.45
BB	0.00	0.00
BC	0.00	0.00
CC	0.06	0.21

susceptibility systems, have been fitted to the data from the maternal cohort study assuming that all recruited animals were subjected to an infection hazard that only depends on birth cohort (Ferguson *et al.*, 1997b). In the absence of any genetic data on the recruited animals, it is only possible to use these models to explore the range of genetic frequencies and the type of genetic control that are consistent maternal cohort study data.

Although the relative risk observed in the maternal cohort study was only 5.14 (95% CI: 2.29,11.56), the ratio of the susceptibilities, s, of the high- and low-susceptibility animals must be at least 20 to explain the observed data for a simple single-locus two-allele system (Ferguson *et al.*, 1997b). This discrepancy arises because first, dams of the study animals will include some low-susceptibility animals that were observed to be cases and some high-susceptibility animals that were not, and second, assuming the bulls are randomly selected with regard to susceptibility class, the maternally exposed and control offspring will be even more similar genetically than their dams were.

The behaviour of single-locus two-allele models was examined over the entire range of genotypic frequencies and susceptibilities that were consistent with the maternal cohort study data. The sole study to date to find any differences in the PrP gene between BSE-affected and BSE-unaffected cattle (Table 7.1) (Neibergs *et al.*, 1994), suggested that susceptibility was controlled by a single locus but with three alleles. Fitting to the data from Neibergs *et al.* (1994) and the maternal cohort study simultaneously, we estimated population allelic frequencies and genotype

Table 7.2 *Maximum likelihood estimates (and 95% confidence intervals) from a model in which a single locus with three alleles determines susceptibility to infection fitted simultaneously to the genotype frequencies observed by Neibergs* et al. *(1994) and the risks of BSE observed in the two arms of the maternal cohort study.*

Genotypes	s_k	Estimated genotype frequencies	
		Confirmed BSE	Unaffected cattle
AA	1.0	0.46	0.28
		(0.30,0.59)	(0.16,0.43)
AB	4.80	0.18	0.022
	(0.94,12.7)	(0.04,0.38)	(0.005,0.059)
AC	0.39	0.31	0.48
	(0.11,1.1)	(0.11,0.55)	(0.42,0.49)
BB	0.0	0.00	0.00
	(0.0,49.6)	(0.00,0.035)	(0.00,0.0035)
BC	0.0	0.00	0.019
	(0.0,4.1)	(0.00,0.11)	(0.00,0.053)
CC	0.16	0.053	0.20
	(0.006,0.82)	(0.00,0.22)	(0.10,0.33)

Histopathological BSE prevalence in the maternal cohort study

	Observed	Estimated
Maternally exposed	0.140	0.130 (0.075,0.202)
Control	0.043	0.053 (0.029, 0.085)

susceptibilities assuming a single-locus susceptibility system with three alleles (Table 7.2).

Although the data were well fit by the model (likelihood ratio goodness-of-fit statistic $X_4^2 = 3.48$, $p = 0.48$), caution is required in the interpretation of these results. The genotypic frequencies were measured using single-strand conformational polymorphism techniques, and no formal sequencing or pedigree analysis has been performed to confirm these results. Clearly, a genotypic analysis of biopsy material from animals in the cohort study would provide extremely valuable data for further analysis of genetic models.

These analyses ignored the data on the time from the birth of the calf to the onset of clinical signs of BSE in the dam that is critical to distinguishing direct maternal transmission from genetically determined susceptibility to BSE infection. Although virtually all of the animals in the cohort study were born less than 250 days before clinical onset in the dam, this range, albeit limited, is sufficient to detect a relationship between the observed level of maternal risk enhancement and the time from calving to clinical onset in the dam.

The mean excess risk of BSE in maternally exposed animals was shown to depend on dam incubation stage (Donnelly *et al.*, 1997c). Similarly, Fisher's exact tests revealed that cohort study pair outcomes depended on the time from birth of the calf to the onset of clinical signs of BSE in the dam (Donnelly *et al.*, 1997c). While such analyses suggest that maternal transmission contributes at least in part to the observed maternally enhanced risk, they do not yield estimates of the parameters of interest.

The models developed in Chapter 6 allow the rate of maternal transmission as a function of maternal incubation stage and level of genetically enhanced susceptibility among the offspring of affected dams to be estimated simultaneously. The maternal relative infectiousness function is assumed to be a step function so that if animals are only infectious for time ω before onset, then

$$\Omega_M(v) = \left\{ \begin{array}{ll} 1 & v \leq \omega \\ 0 & v > \omega \end{array} \right. .$$

Rather than assuming that the risk of feed-borne infection depends only on birth cohort, we assume that the time- and herd-dependent risk of infection from feed, $r_{Fh}(t)$, is equal to the product $k_h r_F(t)$ where k_h is the mean incidence in the hth herd. The time variation in feed risk in the period July 1987 to December 1989, $r_F(t)$, was fitted by a quadratic polynomial. The per-herd (or per-pair) risk of feed infection could not be estimated from the cohort study data alone since the feed risk for doubly negative pairs would have been estimated to be zero. This is unrealistic, since nearly all natal herds in this study experienced high BSE incidence in the cohorts from which the study animals were drawn.

Assuming that all control animals are in susceptibility class 1 and all maternally exposed animals are in susceptibility class 2,

Table 7.3 *Results of the individual survival model with the best fitting incubation period and age-dependent susceptibility/exposure distributions from the global population model (Chapter 5) Model I is the combined model with both maternal transmission and genetic factors, Model II includes only maternal transmission ($s_2 = 1$), Model III includes maternal transmission for the duration of the maternal incubation period ($\omega \to \infty$) in addition to genetic factors, Model IV includes only genetic factors ($\beta_M = 0$), while Model V includes only maternal transmission for the duration of the maternal incubation period ($\omega \to \infty$, $s_2 = 1$).*

Model	X^2 compared	df	Estimates		
	with I		s_2	β_M	ω (days)
I	—	596	2.71	0.082	88
II	7.1	597	—	0.099	88
III	6.3	597	2.71	0.055	—
IV	33.8	598	4.25	—	—
V	12.5	598	—	0.075	—

individual survival models were fitted allowing for maternal transmission over the last ω days of the maternal incubation period and genetic susceptibility ratio s_2. The model results are shown in Table 7.3 using the parametrizations of $f(u)$ and $g(a)$ that provided the best fit to the population-level incidence data in Great Britain (Chapter 5). The model with late-stage maternal transmission and no genetic contribution (II) fits significantly better than the model with no maternal transmission (IV) as well as the model with maternal transmission throughout the maternal incubation period but no genetic contribution (V). However, the combined maternal and genetic model (I) fits significantly better than all other models.

The estimated relative susceptibility in these survival models is not comparable, as it stands, to the genetically linked susceptibilities as presented in Table 7.2, since the dams of the study animals will include some low-susceptibility animals that were observed to be cases and some high-susceptibility animals that were not, and the maternally exposed and control offspring will likely be even more similar genetically than their dams were. Assuming a single-locus two-allele system with relative susceptibilities of s in the HH

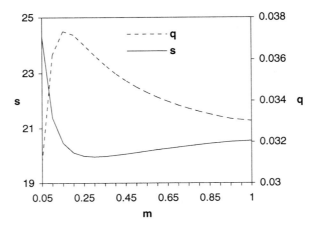

Figure 7.1 *The minimum value of the mean susceptibility in those animals not LL homozygotes, $\bar{s}$, and the corresponding frequency of the H allele, q, that are consistent with the susceptibility ratio of 2.71 (Table 7.3). These results assume that susceptibility to BSE is determined by a single locus with two alleles.*

homozygotes, 1 in the LL homozygotes and $1 + (s-1)m$ in the HL heterozygotes (with $s > 1$), Figure 7.1 presents the minimum value of the mean susceptibility in those animals not LL homozygotes, $\bar{s}$, and the corresponding frequency of the H allele, q, that are consistent with the susceptibility ratio of 2.71. These results demonstrate the relatively high magnitude differences in underlying genetically determined susceptibility levels required to generate the observed ratio.

7.3 Dam−calf pairs in the GB case database

Data on dam−calf links within the epidemiological database allow the assessment of maternal risk enhancement at earlier stages in the maternal incubation period than do data from the maternal cohort study. As a starting point in the analysis of the dam−calf pair data, we compare the number of identified dam−calf positive pairs with that which would be expected by chance, taking proper account of the clustering of cases within holdings. We then analyse the data using individual survival models to estimate the maternal

Table 7.4 *The estimated maternal risk enhancement using pregnancy data where available, assigned weights from dam ages 560 − 730 days and dam and BAB censoring with 95% bootstrap confidence intervals using $\hat{R}$, $\tilde{R}$ and the mean of the within-holding ratios by birth cohort, $\bar{r}$.*

	Birth after	Months from birth of calf until dam onset			
	dam onset	0 to 12	12 to 24	24 to 36	36 to 48
$\hat{R}$	2.77	1.37	1.21	0.86	0.65
	(1.48,4.25)	(1.15,1.59)	(1.04,1.39)	(0.72,1.00)	(0.50,0.81)
$\tilde{R}$	2.84	1.42	1.25	0.88	0.67
	(1.52,4.34)	(1.19,1.66)	(1.08,1.44)	(0.74,1.03)	(0.52,0.83)
$\bar{r}$	2.74	1.62	1.42	0.97	0.89
	(1.35,4.41)	(1.22,2.06)	(1.13,1.73)	(0.76,1.20)	(0.62,1.20)

transmission rate as a function of maternal incubation stage and the magnitude of any genetically enhanced susceptibility in the offspring of BSE-affected dams (Donnelly *et al.*, 1997a).

The risk in calves born to affected dams is found to be significantly enhanced within 24 months of the onset of clinical signs in the dam (Table 7.4). As the period between birth of the calf and onset of clinical signs in the dam increases, the risk enhancement decreases. This pattern is inconsistent with the hypothesis that the maternally enhanced risk results purely from genetically enhanced susceptibility, and therefore provides evidence for direct maternal transmission of the aetiological agent.

Although the consistency between the estimates, $\hat{R}$ and $\tilde{R}$, is reassuring, suggesting that the identification of dams is independent of the disease status of the dam, the significantly *decreased* risk of BSE in the offspring born between 36 and 48 months before the onset of clinical signs in the dam is not explained by either the hypothesis of direct maternal transmission or that of genetically enhanced susceptibility. In contrast, the within-holding estimator, $\bar{r}$, gives nonsignificant, and somewhat greater, estimates for the maternal risk enhancement in such calves. Clustering of dam–calf positive pairs could give rise to such differences between the population-based estimators and that calculated within holdings suggesting that $\bar{r}$ gives more robust estimates.

To fit the models developed in Section 6.4, the ratio of BSE cases to infections, α, must be estimated. Since we analyse the number of BAB cases arising within 4 years of the end of their birth cohort interval (for example 30 June 1989 for the 1989 cohort), the average BAB would be censored at 4.5 years of age if calving were uniform throughout the year. However, since calving is more likely early in the birth cohort interval (in the autumn months), on average BABs are censored between 4.5 and 5 years. Interpolating between the estimated ratios of infections to cases arising by 4 and 5 years of age, obtained from the best fitting population model (Chapter 5) yields estimates of α in the range $18-25$ for cases arising by 4.75 years of age.

Assuming a simple susceptibility class structure with all unaffected dams and their offspring in susceptibility class 1 with $s_1 = 1$ (Section 4.8) and all BSE-affected dams and their offspring in susceptibility class 2, estimates of s_2 and β_{Mk} for all k were obtained using two values of α (18 and 25), $\gamma = 1$ (no underestimation of holding sizes) and $\lambda = 1$ (identical survival probabilities for calves born to BSE-affected and unaffected dams) (Table 7.5). The resulting estimate of s_2 is significantly less than unity since the observed to expected ratios are less than one for the calves born early in the dam incubation period.

Lower survival probabilities for BABs born to BSE-affected dams compared with those born to unaffected dams, $\lambda < 1$, is one mechanism that could account for such a finding. The estimates of s_2 and the maternal transmission rate for calves born after onset in the dam increase dramatically with decreasing λ. for α values consistent with the population model (Figure 7.2 presents results for $\alpha = 18$). The estimates of the maternal transmission rates for animals born 0 to 12 months and 12 to 24 months before dam onset increase more slowly with decreasing λ. While the estimates of the maternal transmission rates for animals born more than 24 months before dam onset decrease for $\lambda < 0.4$, they do not differ significantly from 0 for any values of λ examined.

Since ratios below 1 are inconsistent with both the direct maternal transmission and genetic susceptibility hypotheses, we also present in Table 7.5 the maximum likelihood estimates obtained when $\lambda = 0.9$, the largest value of λ for which all ratios are greater than or approximately equal to 1. The estimated rates of maternal

Table 7.5 *Maximum likelihood estimates for s_2 and β_{Mk} assuming $\gamma = 1$ with the results of the likelihood ratio tests of the null hypotheses H_0 : $s_2 = 1$ and $H_0 : \beta_{Mk} = 0$ for all k and the goodness-of-fit test. The ratios of the expected number of dam–calf BSE pairs from the model (assuming $\lambda = 1$) to the expected number under the assumption of independence between cases of disease in the dams and calves as a function of dam incubation stage are also given.*

| | $\alpha = 18$ | | $\alpha = 25$ | |
	$\lambda = 1$	$\lambda = 0.9$	$\lambda = 1$	$\lambda = 0.9$
s_2	0.825	0.935	0.827	0.945
$H_0 : s_2 = 1$				
p-value	0.016	0.417	0.023	0.517
β_{Mk}				
birth after onset	0.085	0.093	0.117	0.129
0-12 mbo	0.019	0.020	0.026	0.028
12-24 mbo	0.017	0.018	0.023	0.025
24-36 mbo	0.006	0.007	0.009	0.009
36-48 mbo	0.004	0.005	0.006	0.006
$H_0 : \beta_{Mk} = 0 \ \forall k$				
p-value	< 0.001	< 0.001	< 0.001	< 0.001
Goodness-of-fit				
p-value	0.019	0.019	0.019	0.022
Ratios				
birth after onset	2.34	2.55	2.33	2.53
0-12 mbo	1.20	1.31	1.20	1.31
12-24 mbo	1.10	1.20	1.10	1.20
24-36 mbo	0.93	1.02	0.94	1.03
36-48 mbo	0.90	0.99	0.91	1.00

(mbo = months before onset)

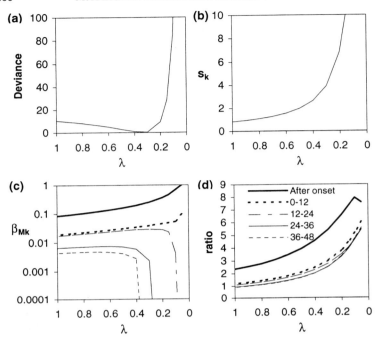

Figure 7.2 *The deviance (a), estimated values of s_2 (b) and β_{Mk} (c) and the ratios of the expected number of dam−calf BSE pairs from the model to the expected number under the assumption of independence between cases of disease in the dams and calves (d) as a function of dam incubation stage and λ assuming $\alpha = 18$.*

transmission, β_{Mk}, are slightly greater than those obtained when $\lambda = 1$ (Figure 7.3), and the estimate of s_2 is no longer significantly below 1.

Although the dam−calf data provide strong evidence of maternal transmission in the late stages of the maternal incubation period, they do not provide evidence of genetically enhanced susceptibility to BSE infection. Indeed the current lack of any evidence for a significantly enhanced risk for animals born more than 2 years before dam onset argues strongly against a genetic component. However, in the absence of pedigree and genotype data the possibility cannot be completely excluded.

The maximum likelihood estimate of s_2 was shown to increase

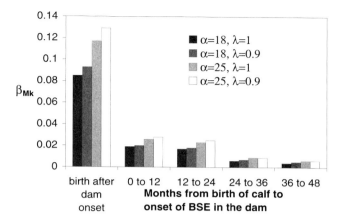

Figure 7.3 *The estimated maternal transmission rates, β_{Mk}, as a function of α and λ.*

linearly with γ (Donnelly *et al.*, 1997a) indicating that the estimate of s_2 obtained from the dam–calf pair data would be closer to that estimated in the analysis of the maternal cohort study if estimated holding sizes were increased. Insight into this relationship can be gained by noting that taking a first-order approximation to $e^{r_{Fhi}s_2}$, ζ_{hi} (6.6) depends linearly on the ratio s_2/γ:

$$\zeta_{hi}\left(x_m|D_{hi}^{-}(r), D_{hik}^{+}(r)\ \forall k\right)$$

$$\simeq \begin{cases} \dfrac{S_{hi}^{+}D_{hik}^{+}(r)(\beta_{Mk}+r_{Fhi}s_2)}{S_{hi}D_{hi}^{-}(r)r_{Fhi}+S_{hi}^{+}\sum_{k=0}^{K}D_{hik}^{+}(r)(\beta_{Mk}+r_{Fhi}s_2)} \\ \qquad\qquad \text{for } x_m = 0,\ldots,K \\ 1-\sum_{k=0}^{K}\dfrac{S_{hi}^{+}D_{hik}^{+}(r)(\beta_{Mk}+r_{Fhi}s_2)}{S_{hi}D_{hi}^{-}(r)r_{Fhi}+S_{hi}^{+}\sum_{k=0}^{K}D_{hik}^{+}(r)(\beta_{Mk}+r_{Fhi}s_2)} \\ \qquad\qquad \text{for } x_m = -1 \end{cases} \quad (7.1)$$

$$\simeq \begin{cases} \dfrac{s_2}{\gamma}\dfrac{S_{hi}^{+}D_{hik}^{+}}{S_{hi}D_{hi}} & \text{for } x_m = 0,\ldots,K \\ 1-\dfrac{s_2}{\gamma}\sum_{k=0}^{K}\dfrac{S_{hi}^{+}D_{hik}^{+}}{S_{hi}D_{hi}} & \text{for } x_m = -1 \end{cases} \quad (7.2)$$

where (7.2) holds for $\gamma D_{hi} \gg \sum_{k=0}^{K} D_{hik}^{+}$ for all i and j and $\beta_{Mk} \ll r_{Fhi}s_2$. For most animals, the latter condition will hold for plausible values of γ since the majority of dams are in an early incubation stage. Although the relative contributions of direct maternal transmission declines with increasing γ, changes in γ have a weaker effect on the values of β_{Mk}

7.4 Suckler calf data

Wilesmith and Ryan (1997) presented data on the fate of the off-spring of BSE-affected pedigree suckler cows in Great Britain. On the basis of their results, they suggested that 'the true risk for off-spring of BSE-affected dams may be less than that indicated by the cohort study.'

The BSE incubation period distribution was estimated from the analysis of the entire BSE epidemic in Great Britain. The probability that onset will have been observed by t years after infection with the aetiological agent of BSE, denoted $F(t)$, was estimated from the maximum likelihood estimate of the best-fitting incubation period distribution (form C in Table 5.1).

Allowing the maternal transmission rate, $\beta_M(u)$, to depend on the time from birth of the calf until clinical onset in the dam, denoted u, the probability that animal i, born u_i before clinical onset in its dam, survives until at least age a_i without experiencing the clinical signs of BSE is thus given by

$$1 - \beta_M(u_i)F(a_i)$$

assuming it was infected at birth. Clearly, this probability would increase if the animal were infected later in life regardless of transmission route.

The probability that no cases of BSE would be observed in the offspring in the follow-up period until August 1996, is thus:

$$p = \prod_i [1 - \beta_M(u_i)F(a_i)].$$

Assuming that $\beta_M(u_i)$ equals β_M for animals born after or within ω months before clinical onset of BSE in the dam, and zero otherwise, an upper 95% confidence limit for β_M can be obtained as a function of ω by solving for β_M setting $\prod_i [1 - \beta_M(u_i)F(a_i)]$ equal to 0.05. The expected number of BSE cases is given by $\prod_i \beta_M(u_i)F(a_i)$.

The p-value and the expected number of cases for the assumed maternal transmission rate of 10% in addition to the 95% confidence interval for β_M are given in Table 7.6 for a range of ω values. The survival ages were reported by Wilesmith and Ryan (1997) in 12-month intervals. In this analysis, the ages were rounded up to

Table 7.6 *The p-value and the expected number of cases for the assumed maternal transmission rate of 10%, in addition to the 95% confidence interval (CI) for maternal transmission rate, β_M, as a function of ω, the number of months prior to clinical onset on the dam at which maternal transmission begins (Donnelly, 1998).*

ω in months	p	Expected cases	95% CI for β_M
2	0.710	0.33	0-0.69
5	0.302	1.15	0-0.24
8	0.225	1.44	0-0.19
11	0.185	1.63	0-0.17
23	0.025	3.58	0-0.08

the nearest year to illustrate the lowest possible bounds that could be placed on maternal transmission on the basis of these data. These results indicate that the data are consistent with a maternal transmission rate of 17.3% and less for up to 11 months before the onset of clinical signs of BSE in the dam. Further, the data are consistent with maternal transmission at a rate of 8.1% for up to 23 months before clinical onset in the dam.

Thus, the data are consistent with maternal transmission over the last 6 months of the maternal incubation period at rates well in excess of that indicated by the cohort study and the analysis of dam−calf pairs in the main BSE database.

7.5 Conclusions

The maternal cohort study, while designed to determine the extent to which maternal transmission of the BSE agent occurs, introduced a number of confounding effects through aspects of its design and implementation. Given this study alone, it is a difficult task to precisely ascribe causes to the observed maternally enhanced risk of BSE infection in the offspring of affected dams. However, the detailed analyses of the cohort data combined with analyses of the dam−calf pair data provide firm conclusions about the levels of maternal transmission.

The results from the full likelihood survival model,

incorporating all available data on the cohort study animals, indicate that the hypothesis of maternal transmission and no genetic susceptibility fit the study results significantly better than the hypothesis of genetic susceptibility and no maternal transmission. These results are robust to model assumptions, and thus give a degree of certainty to the conclusion that direct maternal transmission occurs. The model including both maternal transmission and genetically enhanced susceptibility fits the data significantly better than the hypothesis of maternal transmission alone. However, this result is not robust, losing significance as variations in model assumptions are made. A full exploration of the profile likelihood surfaces for these models suggests a low level of maternal transmission, with rates being highest ($\sim 5-10\%$) for calves born less than 150 days from onset of clinical signs of BSE in the dam.

All models of the dam−calf pair data examined consistently indicate direct maternal transmission of the aetiological agent of BSE in the late stage of BSE incubation (for calves born after or less than 24 months before the onset of clinical signs in the dam) with the risk being greatest for those calves born after dam onset, confirming the findings of the maternal cohort study.

These analyses do not provide consistent evidence of genetically enhanced susceptibility to BSE infection. Indeed, the lack of any evidence in the dam−calf pair data for a significantly enhanced risk for animals born more than 2 years before dam onset argues strongly against a large genetic component. Thus, the results from these analyses suggest the major component of the observed maternally enhanced risk may be attributed to low-level direct maternal transmission in the late stages of the incubation period. However, in the absence of pedigree and genotype data we cannot completely exclude some genetic contribution.

CHAPTER 8

Spatio-temporal correlation and disease clustering

8.1 Introduction

The models and analyses discussed so far have either ignored spatial or holding-level heterogeneity in exposure, or merely controlled for it in order to distinguish differences in risk between classes of animals (*i.e.* when examining maternal risk enhancement). In this chapter we examine the heterogeneity seen in the incidence of reported BSE cases in Great Britain, and derive simple descriptive statistical models to characterize the clustering and temporal correlation seen. We conclude by showing how insight into case clustering can inform the design of efficient culling policies intended to reduce future case incidence of BSE. The understanding of clustering gained in this chapter is then used in the next chapter to construct a detailed stochastic simulation model of BSE transmission dynamics, and to review potential holding-level survival models.

8.2 Spatial structure of the BSE epidemic

Examining the per-capita incidence of BSE through time in England, Wales and Scotland (Figures 8.1 and 8.2), we see that, while there were considerable differences in the scale of the epidemics between these regions, the overall temporal patterns of the epidemics are very similar (Figure 8.3). There is also considerable variation on a county scale (Figure 8.4; for additional years of onset see Anderson *et al.*, 1996).

Since the highest incidence areas are in southern England and the lowest incidence areas are in Scotland, due consideration must be given to the potential for edges to affect measures of spatial structure. We first examine spatial structure of the BSE epidemic looking for overdispersion in a manner which is unaffected by edges but does not take into account the distance between discretized

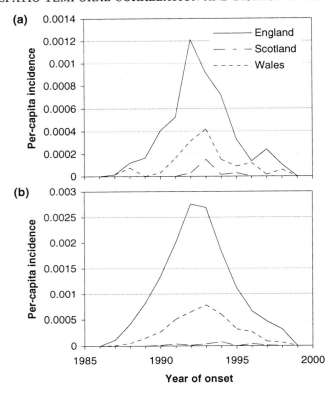

Figure 8.1 *The per-capita incidence of BSE by year of onset in England, Wales and Scotland in holdings of size (a) < 30 cattle and (b) 30–49 cattle.*

spatial units. Having found overdispersion at all scales, we characterize its relationship to spatial distance using more traditional measures of spatial covariance.

8.2.1 Spatial overdispersion in incidence

We can explore how incidence varies with spatial scale in a systematic manner from scales of 100 m to 819 km. Consider a discretization of the land area of Great Britain into M identical squares lying on a regular grid with the size of each square being 100 m × 100 m. We represent the BSE epidemic in this discrete space by defining

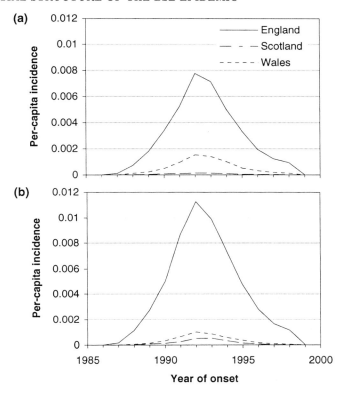

Figure 8.2 *The per-capita incidence of BSE by year of onset in England, Wales and Scotland in holdings of size (a) 50–99 cattle and (b) 100+ cattle.*

the incidence in any square to be the mean per-capita incidence seen in all animals in that square in a particular birth cohort or year of onset. We regard this representation as the saturated model (with M fitted parameters), denoted D_0, of the BSE epidemic at the scale denoted $n = 0$.

We next consider the nested model, D_1, obtained by doubling the width and height of each square, and calculate the incidence in each new larger square as the mean per-capita incidence across all holdings in that square. This new model has $M/4$ parameters, and implicitly assumes there is no significant variation in incidence at a scale less than the size of the new squares − *i.e.* at scales of

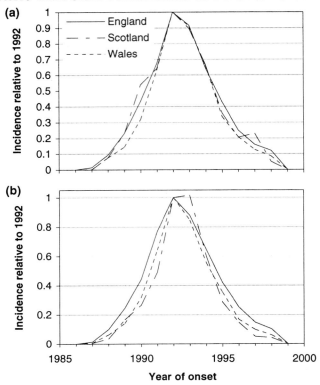

Figure 8.3 *The per-capita incidence of BSE relative to 1992 levels in England, Wales and Scotland in holdings of size (a) 50–99 cattle and (b) 100+ cattle.*

$n < 1$. To test this assumption, we examine the difference in the binomial likelihoods of models D_0 and D_1 to see whether the new model fits the observed epidemic significantly less well (given the reduction in the numbers of parameters fitted) than the previous model assuming that cattle in one square are independent of cattle in all others. The likelihood of model D_1 is simply the saturated likelihood of the data when stratified at the scale $n = 1$ (namely 200 m squares). We can then extend the analysis to arbitrary larger scales n, by considering squares 2^n times the linear size of the smallest square, and denote the resulting model D_n.

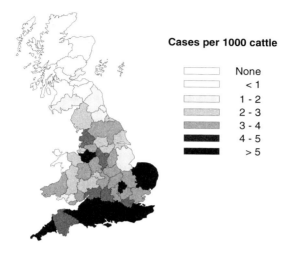

Figure 8.4 *The incidence of BSE in 1992 by county.*

Asymptotically, the likelihood ratio statistic, twice the difference in the log likelihoods of models D_n and D_{n+1}, X_n^2, would be expected to be χ^2 distributed with $3M/4^{n+1}$ degrees of freedom, assuming that cattle in one square are independent of cattle in all others. However, even if the independence assumption holds, the fact that many squares at the 100 m scale have no cases, and some no cattle, means that such asymptotic results cannot be relied upon. It is therefore more robust to perform a bootstrap analysis. We bootstrap the model D_{n+1} to generate a simulated data set of observed incidences for all scale n squares under the null hypothesis that there is no significant variation in incidence at scales below $n + 1$. The likelihood ratio statistic comparing models D_n and D_{n+1} is then calculated for each bootstrap sample, generating the distribution of likelihood ratio statistics under the null hypothesis conditional on the observed distribution of cattle and incidence of BSE. We can then compare the observed likelihood ratio statistics to the null distribution obtained from the bootstrap samples to determine whether the hypothesis can be rejected.

The relationship between this approach and information theory is beyond the scope of this book, but in essence we are using the value of the saturated likelihood of the data stratified at scale n as

Table 8.1 *Tests of spatial overdispersion in incidence in holdings with 100 or more cattle in the (a) 1987 and (b) 1991 cohorts.*

	scale	log likelihood	X^2	95% bootstrap bounds
(a)	100 m	-99546	320	(44,71)
	200 m	-99707	693	(95,138)
	400 m	-100053	2226	(339,420)
	800 m	-101166	8505	(1436,1584)
	1.6 km	-105419	16693	(3098,3343)
	3.2 km	-113765	15706	(3037,3299)
	6.4 km	-121619	8665	(1688,1902)
	12.8 km	-125951	3961	(653,786)
	25.6 km	-127932	1731	(201,290)
	51.2 km	-128797	1537	(54,104)
	102.4 km	-129566	1796	(15,44)
	204.8 km	-130464	2714	(4,24)
	409.6 km	-131821	4140	(0.3,10)
	819.2 km	-133890		
(b)	100 m	-22352	93	(11,25)
	200 m	-22399	169	(19,36)
	400 m	-22484	528	(86,120)
	800 m	-22748	1964	(367,438)
	1.6 km	-23730	4455	(994,1110)
	3.2 km	-25957	5370	(1227,1368)
	6.4 km	-28642	3620	(894,1030)
	12.8 km	-30452	1580	(459,567)
	25.6 km	-31242	641	(163,242)
	51.2 km	-31562	339	(48,93)
	102.4 km	-31731	568	(12,40)
	204.8 km	-32016	661	(8,23)
	409.6 km	-32346	775	(0.3,11)
	819.2 km	-32734		

a measure of the information in the spatial pattern of the epidemic at that scale. By examining how the amount of information in the data varies as a function of the scale of stratification, we gain insight into the heterogeneity of case incidence at different spatial scales.

Table 8.1 shows the results of this analysis on the incidence of

BSE in holdings with 100 or more cattle in the 1987 and 1991 cohorts (similar results are obtained for other holding size categories,
for other cohorts and for incidence by year of onset). It can be seen
that there is significant overdispersion in the incidence of BSE on
every scale examined between 100 m and 819 km. However, despite this heterogeneity, the temporal pattern of case incidences in
different regions is very similar. This implies that there must have
been a number of processes, operating at a range of scales, that
affected the degree to which animals were exposed across Great
Britain (e.g. variation in MBM usage and in feed production methods), but that the recycling of infection must have operated at a
large scale in order for the epidemics in different locations to be
synchronized.

8.2.2 Spatial correlation structure

To gain further insight into the spatial heterogeneity of BSE incidence, we examine two statistics: the spatial correlation function,
and the probability that a holding reports one or more cases conditional on there being a holding distance x away that also reports
one or more cases. More formally, let c_{ih} be the number of cases
arising from cohort i in holding h, and $\mathcal{S}_x = \{\{h_1, h_1'\}, \{h_2, h_2'\}, ...,$
$\{h_{n_x}, h_{n_x}'\}\}$ be the set of all n_x pairs of holdings with spatial sep-
aration x (usually discretized on some scale). The spatial cross-
correlation at distance x between the case incidence in the i and i'
cohorts, $\mathcal{R}_{ii'}(x)$ is given by

$$\mathcal{R}_{ii'}(x) = \frac{\sum_{\{h,h'\}\in\mathcal{S}_x} [c_{ih} - \mathrm{E}(c_i)][c_{i'h'} - \mathrm{E}(c_{i'})]}{n_x \sqrt{\mathrm{Var}(c_i)}\sqrt{\mathrm{Var}(c_{i'})}},$$

where the mean and variances of the cohort-specific incidences are
calculated over all unique holdings occurring in $\mathcal{S}_x$. The conditional
probability at distance x, $\mathcal{P}_{ii'}(x)$ is defined as

$$\mathcal{P}_{ii'}(x) = \frac{\sum_{\{h,h'\}\in\mathcal{S}_x} \mathrm{I}(c_{ih})\mathrm{I}(c_{i'h'})}{\sum_{\{h,h'\}\in\mathcal{S}_x} \mathrm{I}(c_{ih})},$$

where $\mathrm{I}(c_{ih})$ equals 1 if $c_{ih} \geq 1$ and zero otherwise.

Figures 8.5 and 8.6 show these statistics as a function of distance
for incidence in the 1987 cohort alone, but stratified by holding size.
Holding-size stratification is necessary because incidence is correlated with holding size irrespective of the location of the holding,

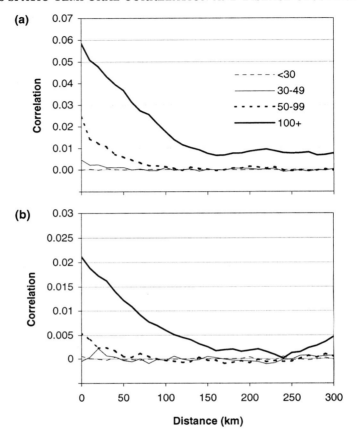

Figure 8.5 *The spatial correlation, $\mathcal{R}(x)$, as a function of distance for the incidence of BSE in the (a) 1987 and (b) 1991 birth cohorts in holdings of < 30 cattle, $30-49$ cattle, $50-99$ cattle and $100+$ cattle.*

but the spatial distribution of holding sizes is not homogeneous across Great Britain (*e.g.* there are many more large holdings in southern England than in Scotland). Both figures indicate significant spatial correlation between case incidence at distances less than about 150 km, and that this correlation is stronger when incidence is higher (as it is for larger holding sizes). We present only results for $x < 300$km, as beyond that distance edge effects

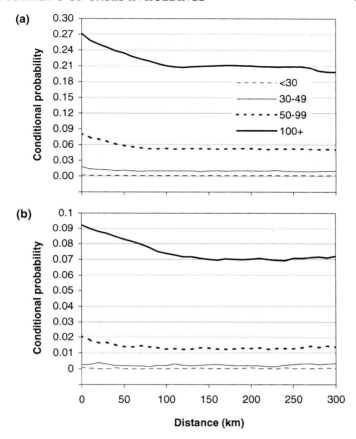

Figure 8.6 *The conditional probability, $\mathcal{P}(x)$, as a function of distance for the incidence of BSE in the (a) 1987 and (b) 1991 birth cohorts in holdings of < 30 cattle, $30-49$ cattle, $50-99$ cattle and $100+$ cattle.*

from the very different incidence levels observed in Scotland and southern England begin to dominate the pattern.

8.3 Clustering of cases in holdings

8.3.1 Observed pattern

One of the key features of the BSE epidemic in Great Britain is the extreme clustering of cases seen, with 20% of holdings

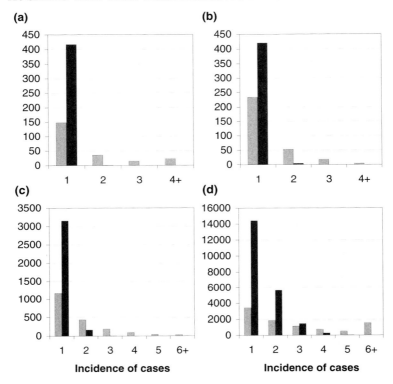

Figure 8.7 *The observed frequency distribution of cases per holding (grey) and the expected distribution assuming a Poisson distribution (black), conditional on the holding having at least one case, for holdings of size (a) < 30 cattle (b) 30–49 cattle (c) 50–99 cattle (d) 100+ cattle. The expected distribution was calculated using the mean number of cases per holding from the entire distribution, including those holdings with no cases.*

reporting over 80% of cases. Under the assumption that exposure to infectious feed was homogeneous across all holdings of equal size, one would expect the resulting incidence of cases per holding to be Poisson distributed. In fact, as Figure 8.7 demonstrates, the observed distributions deviate significantly from Poisson, and are highly over-dispersed.

A useful way of summarizing such clustering is through use of the variance to mean ratio, $\mathcal{K}$, which measures the degree of over

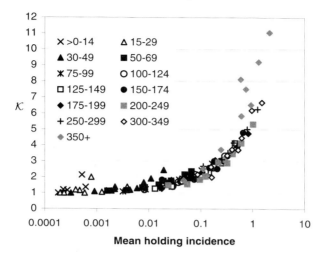

Figure 8.8 *The observed variance to mean ratio, $\mathcal{K}$, of natal holding incidences, stratified by cohort and holding-size category.*

($\mathcal{K} > 1$) or under ($\mathcal{K} < 1$) dispersion of a set of count data. Figure 8.8 shows the value of this statistic against holding incidences for each of the 1985–92 birth cohorts, stratified by holding-size category. Two striking conclusions emerge from examination of this figure: first, an approximately linear increase of clustering with mean incidence, and second, conditional on incidence, there is no apparent effect of holding size on clustering. This second feature – that all the points in Figure 8.8 lie virtually on the same line, regardless of holding size – is especially interesting, given that, as shown in Figure 8.9, mean *per-capita* incidence in Great Britain is seen to increase approximately linearly with holding size and then to remain constant for holdings with 200 or more cattle.

Given the significant clustering seen, it is interesting to ask whether the holdings with highest incidence in one cohort were more highly affected at other times. We therefore consider the correlation within holdings between the case incidence in years i and $i + l$ and between the incidence in birth cohorts j and $j + l$, respectively. Cohort incidence correlation was examined for lags, l, of 1 to 12 years across the 1981 to 1993 birth cohorts, and for lags

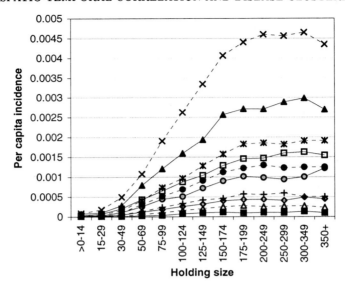

Figure 8.9 *Per-capita incidence by cohort for the 1985–94 cohorts by natal holding size (cohorts in descending order of incidence: 1988, 1987, 1989, 1986, 1990, 1985, 1991, 1992, 1993, 1994).*

up to 9 years when calculating correlation of incidence by year of onset.

The sample correlation functions are displayed in Figure 8.10(a). The plot indicates substantial positive correlation in incidence for lags up to 4 years for cohorts and up to 8 years for years of onset reflecting the clustering of infection within holdings. The higher correlations between years of onset are due to the effect of the incubation period distribution spreading clinical onsets of infections that occurred at one time point over several years. Since one would expect the incidence of cases in successive years of onset to be correlated due to the relatively high variance in the incubation period distribution, even if cohorts were independent, we concentrate on the analysis of correlation in successive birth cohorts for the remainder of this chapter.

Given the intended effect of the ruminant feed ban to interrupt transmission, we might expect the infection process to be more sporadic in cohorts born after the ban, and hence to see less

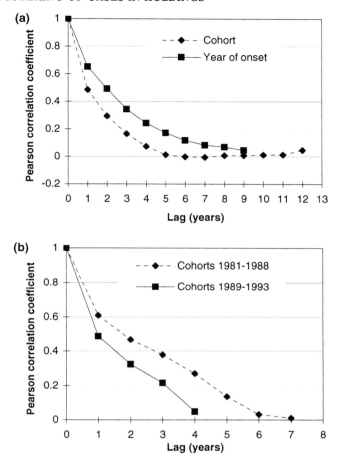

Figure 8.10 *The Pearson correlation coefficient of the incidence within holdings (a) averaged over various cohort and year-of-onset lags and (b) averaged over various cohort lags stratified by cohorts born before and after the introduction of the ruminant feed ban. The corresponding 95% confidence intervals are so narrow as to be invisible on the graph.*

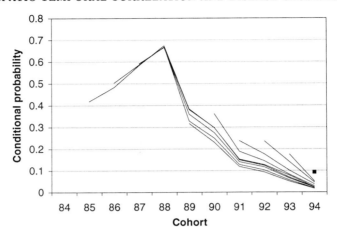

Figure 8.11 *The estimated conditional probability that a holding first affected in a reference cohort between 1984 and 1993 reports one or more cases in cohort x, against x. The probability is calculated conditional on one or more case being reported in preceding cohort y, and probability profiles are shown for all choices of y between 1984 and 1992.*

temporal correlation between incidence in these cohorts. Figure 8.10(b) shows the cohort incidence correlation functions stratified by cohorts born before and after the introduction of the ruminant feed ban, and demonstrates this expected pattern is indeed what was observed.

Further insight into temporal correlation of case incidence can be gained by examining the per-holding probability of reporting at least one case in one cohort conditional on having had a case arise in a preceding cohort. Figure 8.11 shows this probability of seeing one or more cases in a cohort against cohort year, stratified by the cohort being conditioned upon. Interestingly, the profiles obtained do not vary widely as a function of the cohort being conditioned upon until well past the introduction of the ruminant feed ban in mid 1988, indicating again that the mechanisms underlying the observed clustering and temporal correlation were perhaps disrupted by the control measures introduced.

A number of factors may contribute to observed correlation and case clustering. First, although the age-dependent susceptibility/ exposure distribution is tightly peaked, the same batch of infectious feed might simultaneously infect animals of different cohorts. Second, factors that vary between holdings but remain consistent within holdings may contribute to within-holding correlation. For example, holdings that tend to slaughter animals at older ages would, other factors being equal, be expected to have higher disease incidence rates since fewer animals infected at early ages would be expected to be slaughtered prior to the clinical onset of BSE. However, the main contribution to within-holding correlation is likely to be due to variation in husbandry practices between holdings, and in particular to variation in the use (or supply) of protein supplements containing meat and bone-meal. Exploring how such mechanisms can explain the observed patterns of clustering and temporal correlation is the topic of the remainder of this chapter.

8.4 'At risk' holdings model

We begin our more formal analysis of case clustering and correlation by exploring the idea that a subset of all holdings was never exposed to protein supplements containing potentially infectious MBM, and were not therefore at risk of BSE, except through the purchase of cattle infected elsewhere.

The hypothesis that only a fraction of holdings were ever 'at risk' for BSE can be tested with the incidence data on a holding, rather than on an individual animal basis. Assuming that only animals born in 'at risk' holdings could ever have been infected with BSE and that the probability that a specific birth cohort within a holding is affected by BSE is equal for all similarly sized 'at risk' holdings, then we can construct the following model for the first- and last-affected birth cohort in a holding.

We consider C sequential cohorts, $\{1, 2, ..., C\}$, in this analysis. The number of 'at risk' holdings of size j for which i is the *first* cohort affected is denoted n_{Fij}, while we denote the number of holdings affected by BSE in that cohort by n_{ij}. Clearly $n_{F1j} = n_{1j}$. Then, assuming that within any 'at risk' holding there is no correlation between birth cohorts, the expected number of 'at risk' holdings in size category j for which cohort i is the first affected can be written as

$$n_{Fij} = n_{ij} \left(1 - \frac{\sum_{k=1}^{i-1} n_{Fkj}}{N_j^{(AR)}} \right) \qquad (8.1)$$

where $i > 1$ and $N_j^{(AR)}$ is the number of 'at risk' holdings of size j.

Let n_{Lij} denote the number of holdings of size j for which cohort i is the *last* cohort affected. For the final cohort under consideration $(j = C)$, $n_{LCj} = n_{Cj}$. Again assuming that there is no correlation between birth cohorts within 'at risk' holdings, the expected number of 'at risk' holdings in size category j for which cohort i is the last affected can be written as

$$n_{Lij} = n_{ij} \left(1 - \frac{\sum_{k=i+1}^{C} n_{Lkj}}{N_j^{(AR)}} \right) \tag{8.2}$$

where $i < C$.

Under the assumption that all holdings were at risk, $N_j^{(AR)} = N_j$. Assuming only a subset of holdings was at risk and using a model for either first- or last-affected birth cohorts, a maximum likelihood estimate of $N_j^{(AR)}$ can be obtained from the multinomial likelihood where the number of holdings unaffected between cohorts 1 and C is $N_j - \sum_{i=1}^{C} n_{Fij}$ or, equivalently, $N_j - \sum_{i=1}^{C} n_{Lij}$ where N_j is the total number of holdings of size j. Identically structured models can be constructed for the numbers of holdings first and last affected in particular years of onset.

To estimate $N_j^{(AR)}$, we summarized the BSE database to obtain for each holding the birth cohorts affected and the years in which clinical onset of BSE had occurred. For these calculations, animals were linked to their natal holding rather than the holding on which the disease onset occurred, these being different for approximately 33% of BSE cases. Holdings were stratified by holding size (see Figure 8.13 for size categories) and the total number of holdings of each holding size was obtained from the agricultural census data (see Chapter 3).

Figure 8.12(a) shows the expected values of first- and last-affected birth cohorts calculated using (8.1) and (8.2) assuming all holdings are at risk. As we might expect from our previous description of observed correlation between cohort incidences, these models clearly provide poor fits to the data (likelihood ratio goodness-of-fit statistics $X^2 = 22,745$ for first-affected and $X^2 = 21,739$ for last-affected).

Figure 8.12(b) shows how, once the number of holdings at risk by holding size are estimated, the fit improves dramatically (likelihood ratio goodness-of-fit statistics $X^2 = 756$ for first-affected and

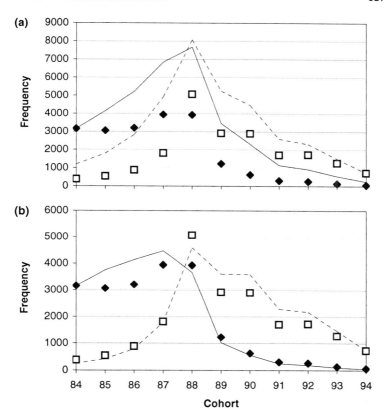

Figure 8.12 *The frequency of first- (diamonds) and last- (squares) affected holdings by cohort for cohorts 1984-94 with fitted values (a) under
the assumption that all holdings are at risk and (b) with estimated numbers of holdings at risk.*

$X^2 = 1000$ for last-affected). The proportion of holdings estimated
to be at risk is seen (Figure 8.13) to increase with increasing holding size, with the estimates being very similar whether estimated
from first- or last-affected holdings on the basis of cohorts or years
of onset. The proportion of 'at risk' holdings ever affected also increases dramatically from 30% for the smallest holdings to 99% for
the largest (Figure 8.13).

Figure 8.14 additionally demonstrates how much better − though

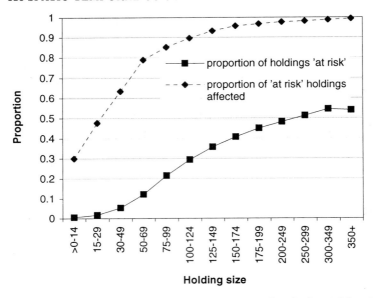

Figure 8.13 *The proportion of holdings estimated to be 'at risk' and the proportion of 'at risk' holdings affected by BSE in any of the cohorts 1984-94 by holding size, as estimated from the analysis of first-affected holdings by cohort.*

not perfectly — this model fits the observed correlation between cohort incidences (as characterized by the conditional probabilities plotted in Figure 8.11) than the model that assumes all holdings were at risk. Furthermore, these results yield further insight into individual risk, as the per-capita incidence, previously noted to increase with increasing holding size (Figure 8.9), is nearly constant when obtained for the 'at risk' holdings only (Figure 8.15), except in the smallest holding size category probably due to poorly estimated holding sizes.

Clearly, additional model detail is still required to obtain a statistically good fit to the data. The model can be expanded in a variety of ways, for example to allow holdings to shift between the 'at risk' and 'not at risk' classes as the epidemic evolved. It is also important to note that these simple models — while explaining most temporal correlation between cohorts — do not explain all the observed within-holding clustering. Figure 8.16 plots the variance to mean ratio, $\mathcal{K}$, of per-holding cohort incidence for holdings

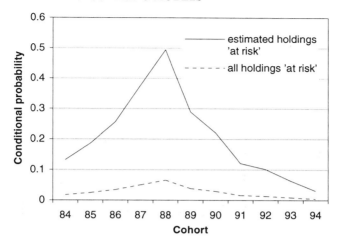

Figure 8.14 *The conditional probability that a holding first affected in a reference cohort between 1984 and 1993 as estimated from the model of first-affected holdings, assuming all holdings are 'at risk' and estimating the number of holdings 'at risk.'*

estimated to be 'at risk,' and shows a very similar pattern of increasing clustering with incidence as is calculated by assuming all holdings are at risk (Figure 8.8).

8.5 Probabilistic clustering models

Here we try to identify the types of exposure/infection heterogeneity that can give rise to the observed (approximately) linear scaling of $\mathcal{K}$ with holding incidence. The basic infection process we will consider is the arrival of an infectious feed unit at a holding. The scale of such a unit might vary from a pellet of feed to a bag or an entire shipment, and we make no assumptions about the proportion of all units that are infectious.

Our model of the effect of an arrival of an infectious feed unit at a holding is that all animals potentially share the unit, with the probability of infection for each animal being p. The number of animals infected by a single unit is therefore binomially distributed with mean Np, where N is the number of animals in the holding.

We then need the probability distribution for the number of animals infected in the holding by the arrival of m separate units.

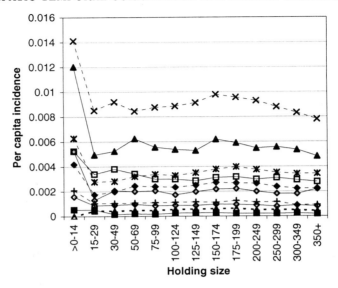

Figure 8.15 *Per-capita incidence by cohort for the 1985–94 cohorts by natal holding size among those holdings estimated to be 'at risk' (cohorts in descending order of incidence: 1988, 1987, 1989, 1986, 1990, 1985, 1991, 1992, 1993, 1994).*

This is simply derived by considering that if $1 - p$ is the probability that an animal escapes infection when exposed to one unit, then $(1-p)^m$ is the probability of escaping infection when exposed to m separate units. Hence p_m, the probability of infection when exposed to m units, is just given by

$$p_m = 1 - (1 - p)^m$$

and the number of animals in a holding infected after exposure to m units is drawn from a binomial distribution with mean Np_m.

We now consider the effect of survival on observed incidence. If s (here $\simeq 0.2$) is the probability that an animal survives long enough to be able to exhibit clinical signs of disease, then sp_m is the probability that an animal will be diagnosed as a BSE case after exposure to m infectious feed units. The number of animals in the holding that are diagnosed as BSE cases after exposure to

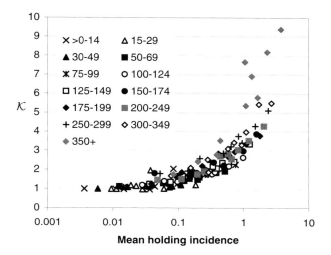

Figure 8.16 *The observed variance to mean ratio, $\mathcal{K}$, of natal holding incidences among the estimated 'at risk' holdings, stratified by cohort and holding size category.*

m units, c_m, is still binomially distributed, and

$$E(c_m) = Ns\left[1 - (1 - p)^m\right]$$
$$E(c_m^2) = Ns\left[1 - (1 - p)^m\right]\left[1 - s + s(1 - p)^m\right] \qquad (8.3)$$
$$+ N^2 s^2 \left[1 - (1 - p)^m\right]^2$$

Alternative infection models can be justified, the most obvious one being that any feed unit is eaten by a fixed number, n_u (which might be 1) of animals in the holding, so that the number infected by one unit is again binomially distributed, but with a maximum of n_u being able to be infected. However, for this type of model, derivation of the probability distribution of the number of BSE cases diagnosed in a holding exposed to m units becomes analytically difficult, and so we do not discuss the topic further.

Given the simple infection model, we now consider the incidence of BSE in specific cohorts, by assuming animals are susceptible/exposed only in the first year of life. The number of infectious units that are consumed by a cohort in one year is then m, and N is the size of the cohort in a holding. To begin with, let us consider the case where the number of infectious units, m, to which a cohort is exposed is drawn from a Poisson distribution with mean λN, where the factor of N indicates that supply of feed units is proportional to holding size (assuming an animal consumes the same volume of feed irrespective of holding size). The first two moments of the number of BSE cases, c, reported from the cohort are, using (8.3) given by

$$E(c) = \sum_m \frac{\exp(-\lambda N)(\lambda N)^m}{m!} Ns \left[1 - (1-p)^m\right]$$

$$E(c^2) = \sum_m \frac{\exp(-\lambda N)(\lambda N)^m}{m!} \left\{ Ns \left[1 - (1-p)^m\right] \right.$$

$$\left. \times \left[1 - s + s(1-p)^m\right] + N^2 s^2 \left[1 - (1-p)^m\right]^2 \right\}$$

One can then show that

$$\begin{aligned} E(c) &= Ns \left[1 - \exp(-\lambda Np)\right] \\ E(c^2) &= Ns[(N-1)s+1] \left[1 - \exp(-\lambda Np)\right] \quad\quad (8.4) \\ &\quad +N(N-1)s^2 \left\{\exp[-\lambda Np(2-p)] - \exp(-\lambda Np)\right\} \end{aligned}$$

and thus

$$\begin{aligned} \mathcal{K} &= \frac{\mathrm{Var}(c)}{E(c)} \\ &= 1 - s + Ns\exp(-\lambda Np) \\ &\quad +(N-1)s\frac{\exp[-\lambda Np(2-p)] - \exp(-\lambda Np)}{1 - \exp(-\lambda Np)} \end{aligned}$$

This is not a particularly transparent expression but can be somewhat simplified if one considers the case where $\lambda Np << 1$ (or equivalently $E(c) << N$, for fixed $s \simeq 0.2$) and performs an expansion to $O(p)$:

$$\mathcal{K} \simeq 1 + (N-1)sp - Nsp\lambda = 1 + (N-1)sp - \frac{E(c)}{N} \quad\quad (8.5)$$

where $E(c) \simeq N^2 s\lambda p$.

We can immediately see from (8.5) that this model can produce significant clustering so long as $(N-1)sp > 1$. Since Nsp is just the average number of cases expected from a holding due to exposure to one feed unit, this condition implies that clustering can occur so long as infectivity is sufficiently aggregated in feed. This model can also reproduce either per-capita scaling of incidence with holding size, if p is assumed to be independent of holding size, since

$E(c)/N \simeq Ns\lambda p$. This assumption implies that cattle feed is very well mixed within a holding, and that there is sufficient infectivity in an infectious unit to be almost certain of infecting any animal consuming even part of the unit. If, however, the results of the 'at risk' holdings model are accepted, then the observed scaling of per-capita incidence with holding size is mostly due to the fact that more large holdings were at risk, in which case assuming p to have the form $p = n_I/N$ for 'at risk' holdings is more reasonable, and gives per-capita incidences that are independent of holding size $(E(c)/N = s\lambda n_I)$, as in Figure 8.15. This would be consistent with the hypothesis that a fixed number of animals are likely to be exposed to any one unit, or that the amount of infectivity in one unit is only sufficient to infect n_I animals.

However, the above model also produces clustering that declines as holding incidence, $E(c)$, rises — diametrically opposed to that observed in the case epidemic. We therefore extend the model by assuming the Poisson rate, λ, at which infectious feed units arrive at a holding is itself drawn from a skewed continuous distribution. For analytical tractability, we assume this distribution is gamma, with PDF:

$$f(\lambda) = \frac{1}{\lambda_0^r \Gamma(r)} \lambda^{r-1} \exp\left(-\frac{\lambda}{\lambda_0}\right) \qquad (8.6)$$

From (8.4) and (8.6), or using the result that a Poisson distribution with a gamma-distributed rate is equivalent to a negative binomial distribution, it is simple to show that:

$$
\begin{aligned}
E(c) &= Ns\left[1 - (1 + \lambda_0 Np)^{-r}\right] \\
E(c^2) &= Ns[(N-1)s + 1]\left[1 - (1 + \lambda_0 Np)^{-r}\right] \\
&\quad + N(N-1)s^2 \left\{[1 + \lambda_0 Np(2-p)]^{-r} - (1 + \lambda_0 Np)^{-r}\right\}
\end{aligned}
$$

and hence

$$
\begin{aligned}
\mathcal{K} &= 1 - s + Ns(1 + \lambda_0 Np)^{-r} \qquad (8.7) \\
&\quad + (N-1)s\frac{[1 + \lambda_0 Np(2-p)]^{-r} - (1 + \lambda_0 Np)^{-r}}{1 - (1 + \lambda_0 Np)^{-r}}
\end{aligned}
$$

Again, a power expansion to first order in p (valid for $\lambda_0 Np \ll 1$) gives

$$\mathcal{K} \simeq 1 + (N-1)sp - \frac{E(c)}{N} + \left(1 - \frac{1}{N}\right)\frac{E(c)}{r} \qquad (8.8)$$

If $N \gg 1$, we can see this gives clustering rising proportional to $E(c)$ with approximate slope $1/r$. For the case where per-capita incidence does not scale with N (*i.e.* for $p = n_I/N$), a more detailed consideration of (8.7) also reveals a slight dependence of $\mathcal{K}$ on holding size, which becomes noticeable for larger $E(c)$. For the case where p is independent of holding size (and per-capita incidence therefore increases linearly with N), a similar pattern is seen, though the $(N - 1)sp$ term in (8.8) needs to be less than unity to avoid $\mathcal{K}$ depending strongly on N at low incidence (which is not seen in the observed pattern of Figure 8.8).

Figure 8.17 shows examples of the clustering generated by this model as a function of incidence and holding size for these two hypotheses about p. It is interesting to note that parameters that give results comparable with the observed patterns (Figures 8.8 and 8.16), correspond to a distribution on the rate of supply of units to holdings that is strongly peaked at zero — meaning most holdings received very few infectious units and only a small minority received many. However, for the number of infectious units (up to 10 per animal per year) received by the latter category to be realistic, the unit size must be assumed to be relatively small ('bag'-sized or smaller) — implying aggregation of infectivity occurred on similarly small scale. Put another way, while the only way the Poisson model could generate clustering was through highly aggregated infectivity (large $(N - 1)sp$, and hence large units), the model with a gamma-distributed Poisson rate (the negative binomial distribution) generates nearly all clustering through skewedness in the exposure distribution, rather than aggregation of infectivity.

Two obvious mechanisms that might have given rise to a highly skewed exposure distribution across holdings are heterogeneity in usage of MBM-containing feed supplements, or heterogeneity in the extent to which different rendering plants and/or feed mills reduced the infectivity in offal from infected animals. For either mechanism, patterns of feed usage or feed-mill processing would have had to remain relatively constant over time (and farmers would have to be relatively loyal to suppliers) to be able to reproduce the observed temporal correlation between cohort incidences. Dynamically generated clustering due to locality of transmission (as seen in directly transmitted infections) is unlikely to have played a very significant role, due to the large spatial scales over which feed distribution occurred and the limited number of feed mills in the United Kingdom.

The next chapter discusses to what extent the mechanisms proposed for generating exposure heterogeneity would be consistent

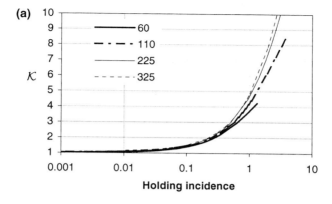

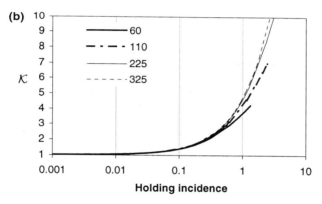

Figure 8.17 *Clustering generated by simple exposure model with gamma-distributed Poisson rate for four holding sizes. Curves were generated by varying λ_0 between 0.001 and 5 (corresponding to differing mean per-animal rates of consumption of infectious units for differing cohorts/herd-size categories), with $s = 0.2$, $r = 0.25$. (a) $p = n_I/N$, where $n_I = 0.12$ and (b) $p = 0.002$.*

with the epidemic nature of BSE (the fact that infectivity was recycled), and their implications for other aspects of the transmission dynamics of the disease.

8.6 Culling policy design

In addition to interventions designed to control the spread of BSE, culling programmes have the potential to reduce the number of BSE cases arising in the future by removing infected cattle from the population prior to the onset of clinical signs of disease. Unless the entire cattle population is to be culled, such programmes should be targeted on those cattle most at risk of infection, and be designed to maximize the number of BSE cases prevented, while minimizing the costs in terms of the number of animals culled and the number of holdings involved. We characterize the effectiveness of a policy by the proportion of future cases prevented and its efficiency by the number of animals culled per case prevented. Subject to cost and implementation constraints, the goal of any cull (indeed of any policy aimed at controlling any infectious disease) must be to jointly maximize both effectiveness and efficiency.

The case clustering discussed earlier in this chapter, together with the temporal correlation of within-holding incidence seen, means that an obvious culling policy option is to target holdings of highest historical BSE incidence. Furthermore, the associations between feed-borne infection risk, cohort of birth and infection status of the dam suggest that policies could be made yet more effective by utilizing these factors when selecting animals to be culled. We therefore explore the following types of culling policies: solely age-targeted − slaughtering all animals born between dates t_0 and t_0'; holding-targeted − slaughtering cattle born in holdings from which at least one case has arisen or alternatively in holdings with greater than a specified per-capita incidence level; maternally targeted − slaughtering animals born within x months of BSE onset in the dam.

Predicting the time-dependent effectiveness of a culling policy requires predictions of future cases as well as estimates of the proportion that will arise from the targeted animals in the absence of culling. To understand how this proportion varies as a function of the number of years since a policy was implemented, the case database was used to explore the effect of a variety of policies if they had been implemented in the past. This form of retrospective simulation of culling policies has the advantage of being model free in its evaluation of relative policy performance. However, to predict future performance of policies, we require predictions of the number of cases expected to arise in future years from both targeted and untargeted animal groups. These are obtained by using predictions of total future case numbers from back-calculation models (see Chapters 4 and 5), and estimating the proportion of those cases that will be prevented by extrapolating the behaviour of retrospectively implemented policies as a function of implementation date into the future.

8.6.1 Age-targeting

Predicting the effect of purely age-targeted policies is relatively easy, given the estimates of cohort specific infection incidence obtained from back-calculation analysis (Chapter 5), as culling a fixed proportion of a cohort at a particular time just prevents the same proportion of cases that would have arisen from that cohort from that time onward.

The efficiency of this type of policy is at its highest if the animals being culled are relatively young, but over 2 years of age. This is because a large proportion of animals are slaughtered at around 2 years, but if culling is left until animals are over, say, 6–7 years of age, a large fraction of cases that might ever arise from the targeted cohort would have occurred already (due to the form of the incubation period distribution).

Table 8.2 shows that had all animals born between October 1990 and June 1993 been culled by 1 January 1997, we estimate that 56% of the cases predicted to arise in the years 1997–2001 would have been prevented. However, the efficiency of this policy, like that of other age-targeted culling policies (Anderson *et al.*, 1996) is very low, compared with policies that take advantage of the significant clustering of cases within holdings by targeting high incidence holdings.

8.6.2 Holding-targeting

We consider a policy that culls all cattle in holdings within a selected cohort from which at least one case had been recorded in the database by a certain time. The effectiveness of such a policy applied to the 1987 or 1988 cohort is shown in Figure 8.18(a), for a range of possible implementation years. Culls are assumed to be implemented at the start of the specified year. A substantial decrease in policy efficiency is seen for later implementation years, since by that time the decline in cases arising from those cohorts

Table 8.2 *Comparison of possible culling policies. A common trend is seen for all the culling policies explored below — that as effectiveness increases, efficiency declines. This is an inevitable result of the strong clustering of BSE cases within holdings and birth cohorts.*

Number	Culling policy description	Effectiveness Cases prevented %	Total cattle culled %	Natal holdings affected %	Efficiency Cattle culled per case prevented
Non-targeted					
1	All cattle	100	100	100	1357
Age-targeted					
2	All cattle born 10/90−6/93	56	22	≤100	523
Holding-targeted					
3	Cattle in cohorts 1990−93 in holdings with one or more cases in the corresponding cohort during 1/91−12/95	24	1.4	5.6	76
4	Cattle in cohorts 1990−92 in holdings with more than one case per 27 cattle in the cohort range during 1/91−12/95	11	0.23	0.57	29
5	As Policy 4 but with a threshold of one case per 50 cattle	22	0.77	1.8	48
Maternally targeted					
6	Cattle born after 10/90 within 6 months of BSE onset in the dam	8.4	< 0.24	≤20	38
Combined maternally and holding-targeted					
7	Policies 4 and 6	19	< 0.47	≤ 23	33
8	Policies 5 and 6	30	< 1.0	≤ 24	45

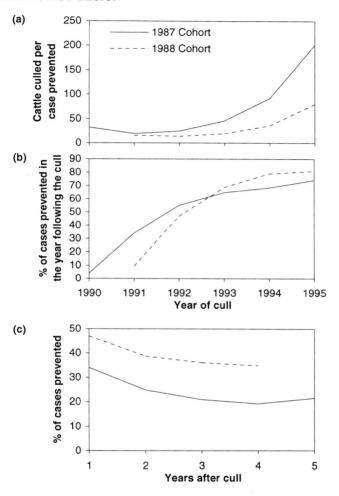

Figure 8.18 *The results of culling policies that cull all cattle in a selected cohort from holdings that had at least one confirmed BSE case in terms of (a) the number of animals culled per case prevented in the year following the cull; (b) the percentage of cases arising from the targeted cohort prevented in the year following the cull; and (c) the percentage of cases arising in the targeted cohort prevented in the years following the cull (assuming the 1987 cohort was culled on 1 January 1991 and the 1988 cohort was culled on 1 January 1992).*

(determined by the incubation period distribution) is considerably faster than the decline in the number of cattle surviving.

Figure 8.18(b) shows the cohort-specific effectiveness of these policies for the year following the cull. The low early effectiveness is due to the fact that when just a few cases have been seen in a particular cohort, only a small number of holdings that will produce cases in that cohort can be identified. The long term effectiveness of such one-off culls is explored in Figure 8.18(c), which shows a relatively slow decline in cohort-specific effectiveness in the years following the cull. The slowness of this decline enhances our confidence in being able to predict the effectiveness of similar culling policies aimed at reducing case numbers from the current time on.

It is this type of policy that was implemented by the British government in early 1997. Cattle targeted were those in the 1990 to 1993 cohorts in holdings from which a case originated in the corresponding cohort over the time period 1991−1995 inclusive. We estimate (Table 8.2) that this policy would prevent 22% of BSE cases predicted to arise in the years 1997−2001. However, it can be seen that while the policy is much more efficient than purely age-targeted culling, it is less efficient than policies that target holdings on the basis of per-capita (rather than case) incidence − since case incidence targeting penalizes larger holdings where the chance of a single case is greater.

The temporal correlation of within-holding incidence shown in Figure 8.10 suggests that holdings that have already experienced a high incidence of infection are more likely to have a high incidence in the future. Culling policies that target high-incidence holdings (those with per-capita incidence above some threshold) are therefore likely to be more efficient than those that target all holdings with non-zero incidence.

Here we consider three incidence thresholds for a cull aimed at one selected cohort (we use the 1988 cohort as an example here): holdings with incidence in the top 20% of all recorded non-zero incidences, those with an incidence greater than one case per 27 cattle, and those with an incidence above one case per 50 cattle.

Figure 8.19 shows that the trends in efficiency and effectiveness with date of implementation are very similar to those seen for case-incidence holding targeted policies (Figure 8.18). However, per-capita incidence targeting gives higher efficiency, but lower effectiveness.

We examine the future effectiveness of two incidence-targeted

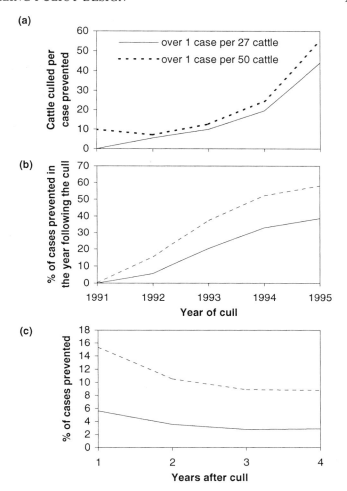

Figure 8.19 *The results of culling policies that cull all cattle in a selected cohort from holdings that experienced incidence greater than a selected threshold in terms of (a) the number of animals culled per case prevented in the year following the cull; (b) the percentage of cases arising from the targeted cohort prevented in the year following the cull; and (c) the percentage of cases arising in the targeted cohort prevented in the years following the cull (assuming the 1988 cohort was culled on 1 January 1992).*

policies (Table 8.2). Effectiveness was estimated by extrapolating curves of the type shown in Figure 8.19(b),(c) and using the resulting cohort-specific effectiveness estimates to modify the case predictions made using the back-calculation model. Culling cattle on 1 January 1997 in the 1990−92 cohorts in holdings from which more than one case per 27 cattle originated in the cohort range in the time period 1991−1995 (inclusive) is predicted to prevent 11% of cases predicted to arise in the years 1997−2001. The same policy but with an incidence threshold of 1 case per 50 cattle is predicted to prevent 22% of cases over the same period, but at the cost of lowered efficiency − though the efficiency is still greater than the government case-targeted policy, which prevents a similar number of cases.

8.6.3 Maternally targeted and combined policies

Evaluation of the effectiveness of maternally targeted policies in the case database requires information linking dams and their calves. While some such information is available for cases born after the 1988 ruminant feed ban, it is insufficient to accurately describe the past effectiveness of maternally targeted policies. However, the estimates of maternal transmission (see Chapter 7) allow prediction of the future effectiveness of these policies. Policies that combine maternal targeting with a per-capita incidence targeting are predicted to achieve particularly high levels of efficiency and effectiveness, in terms of the number of cattle culled per case prevented and the number of cases prevented (Table 8.2).

8.7 Conclusions

In this chapter, we have examined the spatio-temporal correlation structure of the BSE epidemic in Great Britain, and examined simple models that can reproduce the observed pattern of within-holding case clustering. While the BSE epidemic occurred largely synchronously across the entire country, there was substantial overdispersion in incidence of BSE cases at all scales between 100 m and 819 km with stronger spatial correlation apparent in larger holdings, which have on average higher per-capita incidences. Within-holding cohort incidences were significantly temporally correlated for up to 5 years.

The precise causes of this spatio-temporal structure are not yet clear, but the observed temporal correlation, scaling of per-capita incidence with holding-size, and within-holding clustering could be largely reproduced by a combined model that assumed only a subset of all holdings were ever at risk of exposure to BSE, and that exposure within this subset was highly heterogeneous — with most holdings having a low infection risk, and a small minority having much higher risk. Two possible mechanisms that might generate such heterogeneity are variability in husbandry practices (and thus MBM usage), and variability in offal/feed processing (which could generate heterogeneity in the infectiousness of feed). These mechanisms are modelled in greater detail in the next chapter, which also reviews potential holding-level survival models.

While patterns of spatio-temporal correlation and clustering are of theoretical interest to statisticians and modellers (especially in a dataset as large as the BSE case database), we have also shown in this chapter that understanding these patterns is key to the design of efficient culling policies intended to reduce future case incidence. Temporally consistent clustering obviously makes targeting holdings with historically high incidence an attractive prospect for a limited selective cull. However, the effect of clustering is also to cause policy efficiency to fall as effectiveness (the proportion of future cases prevented) rises. Eliminating all future BSE cases is therefore an impossible goal (in the absence of an inexpensive diagnostic test), as it would require slaughter of the entire national holding.

CHAPTER 9

Metapopulation models

9.1 Introduction

The previous chapter described the spatio-temporal correlation and clustering of holding incidences seen in the BSE epidemic, and introduced some simple statistical models that reproduce the observed pattern. However, the clustering models introduced were not suitable for formal parameter estimation, and did not give explicit insight into the potential mechanisms that might generate clustering. This chapter therefore introduces more complex models that begin to resolve these two issues. In the first part of the chapter we propose a survival analysis framework designed to allow maximum likelihood estimation of holding-level infection hazards through time, and briefly discuss the computational and mathematical difficulties posed by such approaches. In the second part, we describe a detailed stochastic simulation model of BSE transmission in multiple holdings that incorporates several mechanisms capable of generating exposure heterogeneity.

9.2 Holding-level survival models

Here we briefly discuss potential approaches to extending the earlier back-calculation analyses to a framework that models the observed clustering of BSE cases at the holding level. It should be noted that the implementation of the type of model discussed here has proved computationally infeasible to date, due to the volume of data being analysed, model complexity, and the larger number of parameters that need to be estimated. Also, while the models outlined below have the benefit of exploring clustering and spatial correlations within a robust estimation framework, it is arguable that the results produced might not give significantly better insight into epidemiological pattern than the simple models detailed in the previous chapter − at least sufficient to warrant the currently huge expenditure of computing resources required. However, as computing power increases in the future, this cost-benefit calculation may change.

9.2.1 Basic formulation

We will assume that cattle in the same holding are independent conditional on the mean feed risk for the holding. However, initially, we will not require that all animals within the same holding have the same feed risk.

Let $p_i[a_{hj}|r_{Fhj}(t)]$ be the probability that the jth animal in holding h that arose from the ith cohort experiences disease onset at age a_{hj} for $a_{hj} = -1, 0, 1, ..., a_{\max_i}$ where $a_{hj} = -1$ indicates that the animal was not recorded to have experienced disease onset, and $a_{\max_i}$ is the maximum observable age at onset for the ith cohort. Let $r_{Fhj}(t)$ denote the feed infection hazard for animal j in holding h at time t. The evaluation of the probability $p_{ihj}[a_{hj}|r_{Fhj}(t)]$ is calculated thus

$$p_i[a_{hj}|r_{Fhj}(t)] = \frac{\int_{t_0=i}^{i+1} \int_{a'=a_{hj}-1}^{a_{hj}} B(t+0)\phi_{RC}^{hj}(t_0,a')da'dt_0}{\int_{t_0=i}^{i+1} B(t+0)dt_0}$$

where $\phi_{RC}^{hj}(t_0,a')$ is the PDF that animal j in holding h born at time t_0 is reported as a case at age a', and we have assumed that age at onset, a_{hj}, and birth cohort, i are stratified in yearly steps. $\phi_{RC}^{hj}(t_0,a')$ is calculated from equation 4.26, with the modification that feed risk is allowed to vary on a per-animal, per-holding basis.

It is clearly impossible to estimate individual feed hazards, but we still wish to be able to capture between-animal heterogeneity in feed risk. We therefore assume that the $r_{Fhj}(t)$ arise from a distribution on the space, $\mathcal{M}$, of all possible such functions (non-negative univariate integrable functions defined on the interval $[t_{\min}, t_{\max}]$). Let $w[r_{Fhj}(t)]$ be such a distribution with mean $r_{Fh}(t)$ for holding h.

The data likelihood for holding h conditional on $r_{Fh}(t)$ is given by

$$L_h = \prod_{i,j} \int_{\mathcal{M}} p_i[a_{hj}|r_{Fhj}(t)]w[r_{Fhj}(t)|r_{Fh}(t)]\mathcal{D}r_{Fhj}(t)$$

where $\mathcal{D}r_{Fhj}(t)$ represents functional integration on $\mathcal{M}$ (effectively a sum over all possible forms of the feed hazard). The conditional likelihood for all holdings is then the product

$$L = \prod_h \left[\prod_{i,j} \int_{\mathcal{M}} p_i[a_{hj}|r_{Fhj}(t)]w[r_{Fhj}(t)|r_{Fh}(t)]\mathcal{D}r_{Fhj}(t) \right].$$

Maximum likelihood estimation of this model now requires the estimation of each mean holding feed risk $r_{Fh}(t)$ in addition to the other parameters. We can simplify further by assuming that the $r_{Fh}(t)$ themselves arise from a distribution on $\mathcal{M}$, $v[r_{Fh}(t)]$. The likelihood for all holdings can then be written

$$L = \prod_h \int_{\mathcal{M}} \left[\prod_{i,j} \int_{\mathcal{M}} p_i[a_{hj}|r_{Fhj}(t)]w[r_{Fhj}(t)|r_{Fh}(t)]\mathcal{D}r_{Fhj}(t) \right]$$
$$\times v[r_{Fh}(t)|r_F(t)]\mathcal{D}r_{Fh}(t).$$

Thus, it is only when all $r_{Fhj}(t)$ are assumed to equal $r_F(t)$ and all animals are independent that the likelihood for the ith cohort can be written as the product

$$\prod_{i,h,j} p_i[a_{hj}|r_F(t)]$$

as in Chapter 4.

In general, parametrization of w and v and evaluation of the functional integral is highly non-trivial. It is therefore necessary to restrict the space of feed risk functions, $\mathcal{M}$.

9.2.2 Simplifications and application to clustering

To gain insight into the observed clustering of cases within holdings (Donnelly *et al.* 1997), one might initially concentrate on between-holding variability alone, and ignore within-holding variability in the feed risk experienced by individual animals. Hence, we assume $r_{Fhj}(t) = r_{Fh}(t)$ for all j and the conditional likelihood for holding h reduces to

$$L_h = \prod_{ij} p_i[a_{hj}|r_{Fh}(t)].$$

Second, by ignoring high-frequency fluctuations in feed risk (given the high degree of temporal correlation observed within holdings), we can represent $r_{Fh}(t)$ in piecewise linear form with predefined knot locations given by $\{T_1, ..., T_K\}$. $r_{Fh}(t)$ is then represented by the vector of knot amplitudes $\mathbf{r_{Fh}} = (r_{Fh}^1, ..., r_{F_h}^K)$, and $\mathcal{M}$ just reduces to R_K^+. This gives us a concrete representation of the functional measure $\mathcal{D}r_{Fh}$:

$$\mathcal{D}r_{Fh} = \prod_k dr_{Fh}^k.$$

Assuming that all r_{Fh}^k are independent, $v(\mathbf{r_{Fh}})$ becomes a product of the individual density functions for r_{Fh}^k (e.g. Weibull or gamma density functions). However, assuming a non-zero within-holding temporal correlation requires a more complex form for the joint density of the $\mathbf{r_{Fh}}$ parameters.

For example, we can make the assumption that the $\log(r_{Fh}^k)$ values, denoted s_{Fh}^k, are normally distributed and have an exponential correlation model such that the correlation between two knots at distance δt is equal to $\alpha^{\delta t}$. The mean of $\mathbf{s}_{Fh} = (s_{Fh}^1, ..., s_{Fh}^K)'$ is $\mathbf{Z}_h \boldsymbol{\pi}$ where $\boldsymbol{\pi}$ is a $p \times 1$ vector of parameters and $\mathbf{Z}_h$ is $K \times p$ matrix of covariates. The $K \times K$ variance-covariance matrix, $\mathbf{V}_h$, of $\mathbf{s}_{Fh}$ is $\sigma_\alpha^2 \mathbf{r}_h + \sigma_\epsilon^2 \mathbf{I}$ where $\mathbf{r}_h$ is the matrix of temporal correlations with elements of the form $\alpha^{|t_l - t_m|}$, $\mathbf{I}$ is the identity matrix and σ_α^2 and σ_ϵ^2 are the variance components. The joint density of $\mathbf{s}_{Fj}$ is

$$
\begin{aligned}
q(\mathbf{s}_{Fh}) &= (2\pi)^{-\frac{1}{2}p} |\mathbf{V}_h|^{-1/2} \\
&\quad \times \exp\left(-\frac{1}{2}(\mathbf{s}_{Fh} - \mathbf{Z}_j \boldsymbol{\pi})' \mathbf{V}_h^{-1} (\mathbf{s}_{Fh} - \mathbf{Z}_h \boldsymbol{\pi}) \right).
\end{aligned}
$$

Thus, the likelihood can be written

$$L = \prod_h \int_{\mathcal{M}} \left[\prod_{i,j} p_i[a_{hj}| \exp(\mathbf{s}_{Fh})] \right] q(\mathbf{s}_{Fh}) \exp\left(\sum_k s_{Fh}^k \right) \mathcal{D}\mathbf{s}_F.$$

This framework could be extended to incorporate spatial correlation between holdings. The $HK \times HK$ variance-covariance matrix (where H is the number of holdings) then has block diagonal elements $\mathbf{V}_h$ and off-diagonal elements denoting the spatio-temporal correlation structure. The spatial correlation structure could be a function simply of geographic distance or perhaps more appropriately of similarity of feed supplies. Data have been collected (though not released) by CVL staff on proprietary concentrates fed to each BSE case throughout its life by supplier/compounder and name of ration. Assuming the distribution and correlation model described above, the joint density of $\mathbf{s}_F$ could then be written

$$q(\mathbf{s}_F) = (2\pi)^{-\frac{1}{2}p}|\mathbf{V}|^{-1/2} \exp\left[-\frac{1}{2}(\mathbf{s}_F - \mathbf{Z}\boldsymbol{\pi})'\mathbf{V}^{-1}(\mathbf{s}_F - \mathbf{Z}\boldsymbol{\pi})\right]$$

where $\mathbf{Z} = (\mathbf{Z}_1, ..., \mathbf{Z}_H)'$. The likelihood is then given by

$$L = \int_{\mathcal{M}} \left[\prod_{i,j,j} p_i\{a_{hj}|\exp[s_{Fh}(t)]\}\right] q(\mathbf{s}_F) \prod_h \exp\left[\sum_k s_{Fh}^k\right] \mathcal{D}\mathbf{s}_F$$

where the functional measure $\mathcal{D}\mathbf{s}_F$ is given by

$$\mathcal{D}\mathbf{s}_F = \prod_{h,k} ds_{Fh}^k.$$

Of course, other distributions and correlation models for the r_{Fh}^k values could be explored. Given the discussion of clustering patterns in the previous chapter, it would obviously be sensible to explore a model where r_{Fh}^k values were assumed to be gamma distributed (with most holdings experiencing very low feed risk), or to arise from a mixed discrete-continuous distribution (*i.e.* some fraction of holdings never being exposed). However, analysis of the correlation structure of such models is complicated by the difficulty inherent in constructing multi-variate non-normal continuous distributions.

9.3 Stochastic simulation models of the BSE epidemic

While the feed-borne transmission route can essentially be seen as horizontal, the complex and indirect nature of the 'contact' process potentially introduces considerable additional exposure heterogeneity caused by, for instance, aggregation of infectivity within MBM, variation in feed processing methods, spatial locality in the production and distribution of feed, and holding-level variation in husbandry practices or cattle demography.

We formulate a stochastic transmission model that explicitly describes the recycling of animal material. This utilizes epidemiological parameter estimates obtained from back-calculation analyses of the BSE incidence data in Britain (See Chapters 4 and 5) and allows the exploration of a wide range of transmission and mixing parameter combinations that produce results consistent with observed cohort incidences in Great Britain. This model allows us to go beyond the simple analytical analyses of the previous chapter and explore the precise effect of different epidemiological heterogeneities on the transmission dynamics of BSE. The first such heterogeneity examined is aggregation of infectivity in feed. Since this might arise at a variety of scales, ranging from that of the bite (due, for example, to the presence or absence of a fragment of nervous system tissue) to the bag or batch (with batches potentially being very large due to the scale of the feed production process), the model therefore describes feed in terms of discrete units that can be arbitrarily sized to explore the effect of aggregation scale on clustering.

We model feed-borne transmission of BSE through the recycling of animal tissue in the form of MBM. The process can be divided into two epidemiological scales. The between-holding variation is represented by the structuring of the cattle population into holdings that are linked through the use of the same feed producers (Figure 9.1(a)). The conceptual structure is that of a metapopulation or patch model, with holdings representing subpopulations that are weakly coupled together by feed producers. Within-holding dynamics are then governed by demographic, infection and incubation processes (Figure 9.1(b)).

This structure, being a simplified description of the actual feed production and distribution process, allows the effects of holding-level heterogeneity in the use of MBM feed, and differing feed producer 'infectivity' to be examined. By feed producer 'infectivity' we mean the extent to which a particular producer reduces the infectivity of the offal used to make feed during the rendering and production process. However, the primary purpose of this section is to outline how complex simulation models can be constructed, rather than present an exhaustive description of model outputs. In reviewing model results we therefore restrict ourselves to demonstrating how the two forms of heterogeneity described above can reproduce observed patterns of clustering.

9.3.1 Between-holding processes

Recycling of animal tissue in animal feed involves slaughtering, rendering and feed production. In the rendering process, offal from cattle may be mixed with offal from other agricultural animals to produce MBM. The MBM is sold to feed producers, for use as an ingredient for animal feed products.

We explicitly model feed production plants (labelled with index

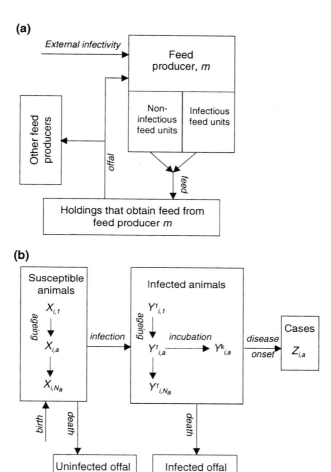

Figure 9.1 *Schematic representation of stochastic simulation model of BSE transmission dynamics (a) between-holding dynamics: the recycling of animal tissue between holdings and feed producers, (b) within holding dynamics: demographic, infection, and incubation processes.*

m) from which individual holdings (labelled with index i) acquire animal feed. Production plants are characterized by two parameters, η_m, reflecting the degree to which the infectiousness of contaminated material is reduced during rendering, and θ_m, reflecting the degree to which one animal's tissue is dispersed over several units of feed. Since we do not explicitly model the rendering process, these two parameters characterize the average MBM supplied by feed production plant m. The effects of disease control measures are modelled by allowing η_m to vary with time.

Animal feed is modelled as discrete units each of which may be infectious. The number of such units consumed per animal per week, c, describes the size of units being used, and hence the scale on which infectivity can aggregate in feed. The relationship between the relative infectiousness of offal $\Phi_m(t)$ and the proportion of feed units infected is given by

$$p_I = 1 - [1 - \Phi_m(t)]^{\theta_m} \qquad (9.1)$$

for $0 \leq \Phi_m(t) \leq 1$, where $\Phi_m(t) = 0$ if there is no infectivity and 1 if all MBM is infectious. The degree of mixing of tissue during rendering and feed production is characterized by θ_m. For small θ_m, infectivity is concentrated in relatively few feed units, while as θ_m becomes large any level of infectivity will infect all feed units.

At each time point, the probability that a holding receives a new delivery of feed is given by

$$\frac{0.6 - \max(q, 0.5)}{0.1}$$

if the holding's store contains less than 60% of the feed required by the holding for six weeks, where q is the proportion of feed remaining in the feed stock. Each feed delivery is the amount of feed required by the holding for six weeks. The number of infectious units in each delivery is a binomial random variable with probability p_I (9.1). Each feed producer begins with a stock of 14 weeks feed for all holdings to cope with chance fluctuations in the demand. Model results are not very sensitive to the choice of the feed delivery parameters. For the parameters used, the mean storage time of feed units is approximately 3 months.

Between-holding interaction is modelled through groups of holdings acquiring feed from the same feed producer. For simplicity, we assume that each holding i only buys feed from feed producer $m(i)$, and mainly consider a system with a single feed producer.

9.3.2 Within-holding processes

Within each holding, animals are stratified by age a (discretized in quarter-years) and incubation stage k. Let $X_{i,a}$, $Y_{i,a}^k$ and $Z_{i,a}$ denote the number of susceptible animals, infected animals in incubation stage k, and clinical cases, respectively, at age a in holding i.

We model a set of holdings comprising four size categories (< 30, $30-49$, $50-99$ and $100+$ adult cattle). Holdings within one category start off being identical replicas. As the stochastic simulation progresses, demographic processes result in variation in holding sizes within each category. Initially, all animals in a holding are susceptible with an age distribution taken from the mean age distribution for demographic equilibrium in the absence of BSE-related mortality (see Section 3.3).

Population dynamics

Deterministically, infection, incubation and demographic processes are governed by the following equations

$$\frac{dX_{i,1}}{dt} = \sum_{a'} \alpha_{a'} N_{i,a'} - Q_{i,1} X_{i,1} - \mu_1 X_{i,1}, \tag{9.2}$$

$$\frac{dX_{i,a}}{dt} = -Q_{i,a} X_{i,a} - \mu_a X_{i,a}, \qquad (1 < a \le N_a), \tag{9.3}$$

$$\frac{dY_{i,a}^1}{dt} = Q_{i,a} X_{i,a} - \mu_a Y_{i,a}^1 \qquad (1 < a \le N_a), \tag{9.4}$$

$$\frac{dY_{i,a}^k}{dt} = -\mu_a Y_{i,a}^k + \nu_{k-1} Y_{i,a}^{k-1} - \nu_k Y_{i,a}^k \quad (1 \le a \le N_a, \tag{9.5}$$
$$1 < k < N_s),$$

$$\frac{dZ_{i,a}}{dt} = \nu_{N_s} Y_{i,a}^{N_s} \tag{9.6}$$

where $N_{i,a} = (X_{i,a} + \sum_k Y_{i,a}^k)$, μ_a is the age-dependent slaughter rate, α_a is the age-dependent calving rate, ν_k is the transition rate from incubation stage k to $k + 1$, and $Q_{i,a}$ denotes the infection hazard experienced by an animal of age a in holding i.

Animals age deterministically, moving from $X_{i,a}$ to $X_{i,a+1}$ every three months. Similarly, to obtain a two-year time delay in the incubation period (see Chapter 5), the first eight quarter-years of the incubation period are updated deterministically along with the aging process. Other processes are implemented stochastically, using a discrete time approximation: *i.e.* the changes in $X_{i,a}, Y_{i,a}^k$, and $Z_{i,a}$ occurring after each (small) time step are drawn from Poisson distributions. The rates governing these processes (birth, slaughter, infection, post 2-year incubation) are given by the relevant terms on the right-hand side of (9.2-9.6).

We use multiple (stochastic) incubation stages to allow the post-2-year incubation period distribution to be non-exponentially distributed (Cox and Miller, 1965, pages 252–271). The transition probabilities for these incubation stages are chosen to be $\nu_k = 0.0356$/week for all k (giving a gamma distribution), so that the overall incubation period distribution closely approximates that estimated using the back-calculation model applied to the incidence data from Great Britain (see Chapter 5). Animals leaving the last incubation stage reach clinical onset of BSE (9.6).

We assume that animals older than 2 years calve annually. Age-dependent slaughter rates are calculated from the survival function described in Section 3.3.

Infectiousness of offal

We assume that the infectivity of offal from animals varies with incubation stage, with the infectiousness of offal from an animal in incubation stage k given by

$$\Omega_k = \zeta + (1 - \zeta) \exp[-bT(k)],$$

where ζ is the baseline infectiousness, b is the growth rate of infectiousness during incubation, and $T(k)$ is the mean time for animals to progress from the k-th incubation stage to disease onset.

The relative infectiousness of all offal at time t for feed producer m is then given by

$$\Phi_m(t) = \sum_k \Omega_k p_m^k(t) + \delta_m(t),$$

where p_m^k is the proportion of animals slaughtered at time t that are in incubation stage k, and $\delta_m(t)$ is an external infectivity source, representing the infectivity of offal unrelated to feed-borne BSE infection (*e.g.* a scrapie-like or sporadic BSE-like agent). Such an external source is required to initiate the simulated epidemics.

We consider two types of external infectivity source, a single event at the start of the simulation, and a low-level continuous source that increases linearly over the first 3 years of the simulation and is constant thereafter.

In the situation where all slaughtered infectious animals have maximal infectiousness and where the relative contribution of δ_m to $\Phi_m(t)$ is marginal, we may think of the tissue–mixing parameter θ_m in (9.1) as the average number of different animals from which the MBM in one unit of feed originates.

The within-holding infection hazard

The dynamics in different holdings are coupled through the infection hazard $Q_{i,a}$, given by

$$Q_{i,a}(t) = g(a)\eta_{m(i)}(t)e_i(t),$$

where $g(a)$ is the degree of susceptibility/exposure of animals of age a (normalized to reach a maximum value of unity) calculated from back-calculation model estimates, $m(i)$ denotes the index of the feed producer that supplies holding i, $\eta_{m(i)}(t)$ is the relative infectiousness of feed units originating from feed producer $m(i)$ (representing the degree to which the production process reduces the infectivity of the raw material), and $e_i(t)$ is the proportion of animals in holding i that are exposed to infectious feed at time t.

When n feed units are delivered to holding i, from which $\tilde{n}$ are infectious, these are converted into n/c weekly portions (of which $\tilde{n}/c$ are infectious) that are added to the farmer's store. As explained above, the parameter c, the number of feed units consumed per animal per unit time (*i.e.* per week), controls the scale on which infectivity can aggregate in feed. The proportion of animals exposed per unit time, $e_i(t)$, is a function of $f_i(t)$, the proportion of feed units consumed that are infectious – assumed to be equal to the proportion of portions in the holding feed store that are infectious at that time.

The mathematical relationship assumed between $e_i(t)$ and $f_i(t)$ determines the manner in which feed units are distributed among animals. We consider two different models for this relationship, one labelled mass-action (MA) and one pseudo-mass-action (PMA) (Anderson and May, 1991; De Jong *et al.*, 1995). For mass-action feed consumption, the animals that are unexposed at time t are those that have not eaten any infectious portions. Hence,

$$e_i(t) = 1 - [1 - f_i(t)]^{\gamma_{MA}} \qquad (9.7)$$

where γ_{MA}, the 'feed consumption' parameter, measures the average number of feed units per week from which an individual animal will eat (allowing for animals to share feed units). For small values of $f_i(t)$, (9.7) may be approximated by

$$e_i(t) \approx \gamma_{MA} f_i(t). \tag{9.8}$$

For the pseudo-mass-action feed-consumption model, the proportion exposed, $e_i(t)$, is taken to be

$$e_i(t) = 1 - (1 - \gamma_{PMA})^{cN_i(t)f_i(t)}. \tag{9.9}$$

In this model, the feed is distributed among the animals in such a way that, for every infectious unit consumed, a proportion γ_{PMA} of the holding is exposed to the risk of infection. For small values of γ_{PMA}, (9.9) may be approximated by

$$e_i(t) \approx \gamma_{PMA} cN_i(t) f_i(t). \tag{9.10}$$

The nomenclature mass-action/pseudo-mass-action derives from simple transmission models (see De Jong et al.(1995)). For (true) mass-action the infection hazard is taken to be proportional to the density or proportion of infectious units (individuals), whereas for pseudo-mass-action it is taken to be proportional to the number of such units (individuals), as seen by comparing (9.8) and (9.10).

9.3.3 Parameter estimation

Latin-hypercube sampling of the space of all transmission parameters can be used to determine parameter sets giving model results consistent with the observed case data. We accept a parameter combination when the simulated incidence in two consecutive cohorts is consistent with the observed 1985 and 1986 cohort incidences on the basis of a likelihood ratio test (with 2 degrees of freedom). Later cohorts are not used as these were potentially affected by the ruminant feed ban introduced in July 1988, while the incidence data from earlier cohorts suffer from under-reporting. Our primary interest is in exploring epidemic patterns that are broadly consistent with the early growth phase of the BSE epidemic, rather than exactly replicating the entire epidemic. The latter exercise would require time-varying infectiousness/consumption parameters, a complexity that is more easily incorporated into back-calculation models (Chapter 4).

The deterministic version of the model is used to test

parameter combinations. While the epidemic produced by a deterministic model is never identical to the mean of all possible epidemic realizations of the equivalent stochastic model (particularly if clustering is significant), the differences during the initial phase of epidemic growth are slight enough to justify the approximation, given the dramatic reduction in the computational burden required to perform extensive parameter sampling.

We do not present detailed results from this sensitivity analysis here, as our principal aim is to explore patterns of case clustering. However, a very broad range of transmission parameters is found to be consistent with the 1985 and 1986 cohort incidences, the primary restriction being that R_0 values had to be broadly consistent with the estimates presented in Chapter 5 − namely over 10 if infectiousness was largely restricted to the end of the incubation period, dropping to below 2 if animals were equally infectious at all stages of BSE incubation. This restriction still permits a wide range of possible case clustering patterns to be generated.

9.3.4 Results: Insight into case clustering

Since clustering is an essentially stochastic process, its exploration requires generation of multiple stochastic realizations of the epidemic model. Simulations utilized a representative subset of the accepted parameter combinations discussed in the previous section. Epidemics are simulated in systems of between 1800 and 3600 holdings (distributed between four size categories in proportion to their respective frequencies calculated from agricultural census data for Great Britain (Section 3.3)), and generally, a single feed supplier is modelled. This distribution of holdings (the maximum computationally possible) forms a representative sample of up to 3% of the national herd in Great Britain.

As would be expected from the discussion in the previous chapter, significant clustering is generated when infectivity is highly aggregated (small c; *i.e.* large feed units), as shown in Figure 9.2, since there is greater between-holding variability in exposure when infectivity enters holdings in large units. Given highly aggregated infectivity, the degree of clustering generated is secondarily sensitive to the relative infectiousness η, small values of η giving rise to weaker clustering.

Figure 9.2 shows the variance to mean ratios, $\mathcal{K} = \sigma^2/\mu$, generated by the pseudo-mass-action model with highly aggregated

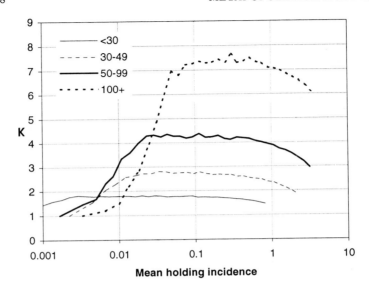

Figure 9.2 *Variance to mean ratio, $\mathcal{K}$, of cohort incidences in four hold-ing size categories (with 3600 holdings simulated in total), averaged across 300 realizations, for pseudo-mass-action model with significant aggregation of infectivity in feed (small c) but no other heterogeneity. Pa-rameters (randomly drawn from combinations consistent with 1985/86 cohort incidences): $\theta = 2.60$, $\eta = 0.855$, $\gamma_{PMA} = 0.830$, $\zeta = 0.0091$, $b = 1.03$ and $c = 0.0046$.*

infectivity in feed, stratified by holding size category. Results were calculated from 300 realizations of a model with 3600 holdings. It should be noted that the fall in $\mathcal{K}$ to 1 for low incidence in Figure 9.2 is a model artifact caused by the finite number of hold-ings modelled. If M holdings are simulated in a size category, then the smallest detectable incidence in any one realization is one case in all the holdings simulated, giving a lower per holding incidence detectability threshold of $1/M$. In general, for a mean per hold-ing incidence C in M holdings, there can be at most CM cases in one herd, and it is simple to demonstrate that this value also represents the absolute maximum detectable level of clustering, $\mathcal{K}$ for this incidence level. Thus, the degree of clustering is increas-ingly underestimated as mean incidence declines. Indeed, once this

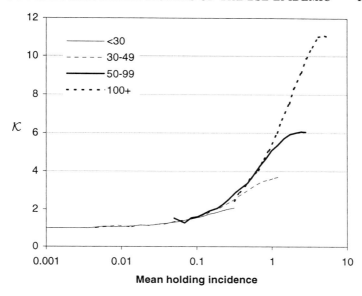

Figure 9.3 *Variance to mean ratio, $\mathcal{K}$, of cohort incidences in four hold-ing size categories (with 1800 holdings simulated in total), averaged across 300 realizations, for pseudo-mass-action model with little aggre-gation of infectivity in feed (large c) but with a highly skewed distribu-tion of holding feed usage (truncated Weibull form used). Parameters (randomly drawn from combinations consistent with 1985/86 cohort in-cidences): $\theta = 1.16$, $\eta = 0.893$, $\gamma_{PMA} = 0.023$, $\zeta = 0.018$, $b = 2.36$ and $c = 1.0$.*

bias is allowed for, Figure 9.2 is in fact consistent with the simple 'Poisson' model of the previous chapter, where $\mathcal{K}$ is expected to be constant for a wide range of incidences above zero, and to fall off only when infection saturation occurs at high incidence. It also demonstrates the scaling of clustering with holding size expected from the pseudo-mass-action model.

As we also might have expected from the results of the simple analytical model presented in the previous chapter, Figure 9.2 does not accurately reproduce the pattern of increasing clustering with incidence that was seen in the case data. On the basis of that anal-ysis, we therefore discard infectivity aggregation as a mechanism for generating clustering, and concentrate on two other forms of

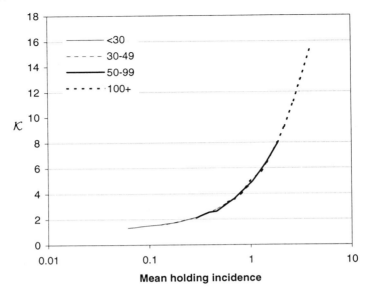

Figure 9.4 *Variance to mean ratio, $\mathcal{K}$, of cohort incidences in four hold-*
ing size categories (with 2400 holdings simulated in total), averaged
across 300 realizations, for mass-action model with little aggregation of
infectivity in feed (large c) but with a highly skewed distribution of feed
producer infectivities (10 mills simulated, with holdings being randomly
assigned to producers at the start of each simulation. Two producers
have 4.5 times the average infectivity, and the remainder 1/8 the aver-
age infectivity). Parameters (randomly drawn from combinations consis-
tent with 1985/86 cohort incidences): $\theta = 3.14$, $\eta = 0.13$, $\gamma_{MA} = 3.26$,
$\zeta = 0.011$, $b = 1.00$ and $c = 0.72$.

heterogeneity. The first is to assume that the demand for MBM-
containing feed varied between holdings, with the distribution of
per-holding feed consumption being highly skewed. It can be seen
from Figure 9.3 (plotted for the pseudo-mass-action model) that
this generates clustering that much more closely resembles the observed pattern. The second heterogeneity is variability in the relative infectiousness (*i.e.* degree to which infectiousness is removed
from offal) of different feed producers. Figure 9.4 shows how in-
troducing this into the simulation model (in this example for the
mass-action model) can also generate very significant clustering.

Although both mass-action and pseudo-mass-action models can generate case clustering within a holding size category, only the pseudo-mass-action model can reproduce the observed linear dependence of per-capita incidence on holding size without assuming, for instance, that larger holdings were more likely to employ more intensive farming methods, and thus use more MBM feed. This is because, for the pseudo-mass-action model, given a sufficiently low infection prevalence, the infection risk experienced by an animal is proportional to the number of infectious feed units from which the animal has eaten (9.10), and it is assumed that feed is sufficiently well mixed within a holding for the probability of consuming a part of any one feed portion to be independent of the size of the holding. Since the total consumption of feed is proportional to holding size, this gives rise to a per-capita infection risk that also scales with holding size. In contrast, for the mass-action model, the per-capita risk is always independent of holding size. However, as discussed in the previous chapter, the observed scaling of per-capita incidence with holding size can also be explained by assuming only a subset of holdings of any particular size was ever 'at risk' of BSE infection – in which case mass-action transmission may have occurred within the 'at risk' subset (Section 8.4).

9.4 Conclusions

In this chapter we have reviewed two more sophisticated modelling frameworks for examining heterogeneity in BSE transmission dynamics – a survival model that takes account of between-holding variation and spatio-temporal correlation in exposure, and a stochastic model that simulates the detailed mechanisms of disease transmission that are thought to have given rise to the BSE epidemic. Optimally, of course, these two approaches would be integrated into a single coherent estimation framework – perhaps based on Markov chain Monte Carlo (Gibson, 1997; Gibson and Renshaw, 1998) or other techniques allowing parameter estimation for stochastic non-linear models. However, since the large volume of BSE incidence data makes fitting the non-stochastic models outlined in section 9.2 computationally unrealistic at present, it will be several years before large-scale integrated analyses become feasible.

Indeed, computational restrictions mean that exploration of the simulation model discussed above has so far been limited to exploration of potential epidemiological mechanisms to generate the observed within-holding clustering of BSE cases. The results reinforce the analysis of the previous chapter, and demonstrate that some combination of a highly over-dispersed distribution of feed usage and variability in the extent to which feed mills removed infectivity from their products is the most likely explanation of observed pattern. In future, the potential exists for the model to give useful insights into mechanisms that might have generated the spatial correlation in case incidence seen in the case data, and perhaps even to shed a little more light on the origins (or at least early growth) of the BSE epidemic in Great Britain.

CHAPTER 10

Predictions and scenario analysis for vCJD

10.1 Introduction

By 31 December 1998, 39 deaths due to a new variant of Creutzfeldt-Jakob Disease (vCJD) in humans had occurred in the United Kingdom. Scientific evidence in support of the hypothesis that the new variant of CJD is a direct consequence of exposure to the aetiological agent of BSE has accumulated since a British government announcement warned of this possibility in March 1996 (Will *et al.*, 1996; Collinge *et al.*, 1996; Hill *et al.*, 1997; Bruce *et al.*, 1997). It is therefore now believed that vCJD arose via exposure to or consumption of products derived from BSE-infected cattle over the period from the early 1980s to mid 1990s. This chapter addresses the question of what information on the potential future course of the vCJD epidemic can be inferred from the limited incidence data available at present (see Figure 3.4).

The work described here was motivated by requests from policy makers (and the public) for robust predictions of the potential scale of the public health threat posed by vCJD − an understandable desire when human communities are faced with a new infectious disease of largely unknown aetiology. The last comparable situation was that faced by statisticians at the start of the HIV epidemic, and in that case, early analyses used a variety of techniques (back-calculation (Brookmeyer and Gail, 1986, 1988; Isham, 1989; Bacchetti, 1993), transmission models (Anderson *et al.*, 1989)) to predict future trends from observed case incidence data. These studies were aided by the partial information on the incubation period distribution (IPD) that was provided by an early cohort study of individuals infected via contaminated blood products (Medley *et al.*, 1987, 1988). However, it should be remembered that it was only when data from large anonymous seroprevalence surveys were incorporated into the analyses that a substantial increase in predictive power is achieved.

The task facing those wishing to predict the scale of any vCJD epidemic is at least as difficult as that posed by early HIV epidemic prediction. The IPD is unknown, and it will be several years before prevalence surveys utilizing highly specific and sensitive tests of pre-clinical infection are completed. Furthermore, one cannot assume that the early pattern in vCJD case numbers would be one of exponential growth — since the majority of any epidemic is likely to have been caused by historical exposure to an exogenous infectious source (meat from BSE-infected cattle), rather than due to horizontal transmission of infection within the human population. The results of back-calculation analyses of the BSE epidemic can provide some insight into the scale (some 467,000 infected cattle were slaughtered for human consumption prior to the introduction of the ban on the use of specified bovine offal in food in November 1989 (the SBO ban; see Bovine Offal (Prohibition) Regulations, 1989) and a further 299,000 afterward (Chapter 5)) and temporal form of such historical exposure, but many key uncertainties remain. These concern the absolute infectiousness of BSE-infected bovine tissue to man, the distribution of infectivity in different tissues as a function of bovine incubation stage, meat product consumption patterns in humans, effectiveness of control measures, and potential genetic variation in human susceptibility to vCJD and in incubation periods.

Given this uncertainty and the limited amount of incidence data available, traditional model fitting techniques are of limited use in exploring potential future trends in vCJD incidence and the epidemiological determinants of epidemic size. Any reasonable model is likely to be over-parametrized, given the data available, and the small number of cases make the confidence region around any estimates so large as to make quotation of such estimates largely meaningless and potentially misleading. This is a particular issue in the context of providing advice to policy makers, where point estimates have a tendency to be seized on, and confidence limits largely ignored.

Similar problems arise when formal risk analysis methodologies (Oliver and Smith, 1990) are adopted. These techniques were developed for assessing failure risks for sensitive engineering projects, and thus tend to be comprehensive in their inclusion of every possible factor that might affect the outcome statistic of interest. In the context of such projects great detail can be justified when, for instance, failure rates of individual system components can easily be estimated. However, the increasing diversity of applications of these techniques raises concerns when similarly detailed event trees, or influence diagrams, are constructed for applications where no reasonable estimates of key parameters are available. In such contexts, practitioners tend to rely on the Bayesian approach of assigning subjective probabilities (judged by expert panel opinion) to a range of parameter values. Given that exact calculation of the joint prior distribution (or posterior distribution, if data are being fitted to) as a function of all model parameters is then frequently precluded by model complexity, extensive parameter sampling should then be performed to determine the sensitivity of risk estimates to uncertainty in model parameters or structure, with the best estimate of risk being that which maximizes the joint prior or posterior distribution. However, in the applications of risk analysis to BSE and vCJD (Comer, 1997), such sensitivity analysis has been woefully inadequate, to the extent that the prior distributions assigned to many parameters were largely ignored in executive summaries of the resulting reports and single 'best guess' values of individual parameters were used to present single 'best estimates' of risk. These reports were then used as the basis of scientific advice to government on issues such as, for instance, banning beef on the bone.

To avoid such pitfalls, we present a model framework of the minimum complexity required to capture the key features of human exposure to BSE and the resulting pattern of any vCJD epidemic. Individual exposure pathways (e.g. via different meat products) are not modelled, but this detail is encapsulated within an overall age- and time-dependent exposure function for the human population. Lack of knowledge of parameters underlying this profile and the incubation period is not then expressed in Bayesian terms, as in many cases there is virtually no basis on which to justify any particular form of prior distribution. Instead, we extensively sample from the entire parameter space to determine the widest range of epidemics that are consistent (at the 95% level) with the time-course and age structure of the vCJD epidemic to date. Our intention is not to be able to assign posterior probabilities to any particular area of parameter space or any range of final epidemic sizes, but to explore the epidemiological determinants of epidemic size and time-course in the most rigorous manner feasible. A further goal is 'what if' scenario analysis – namely investigation of how epidemic predictability may change with time.

The analysis presented here is an extension of that presented in earlier work (Ghani *et al.*, 1998a), and utilizes data from cases reported up to the end of 1998, stratified by age and year of death. Up to that time, cases were only diagnosed following *post-mortem* neuropathological examination. An alternative approach (which produces similar results) is to use dates of clinical onset, but these data are likely to be less reliable due to the subjective (and often retrospective) manner in which such dates are estimated.

We start by outlining the modelling approach and basic results. We then turn to one practical application for this type of analysis – informing the design of large-scale unlinked anonymous testing (UAT) programmes intended to give cross-sectional measures of infection prevalence.

10.2 Survival model

Our aim is to explore the ranges of key epidemiological parameters and sizes of vCJD epidemics that are consistent with the age- and time-stratified incidence data seen to date (Figure 3.4). We therefore construct a survival model describing the probability that an individual with particular characteristics will die due to vCJD infection by a certain time. The model is similar to that developed for the back-calculation analysis of the BSE epidemic, in that we construct an expression for the time-varying hazard of infection, and then convolute this with the IPD pertinent to the type of infection and/or host. However, there is a key difference between this and the BSE back-calculation analysis in the manner in which the models are applied. In the BSE case, our principal goal was to estimate the past pattern of infection hazard with time from a large volume of incidence data. Here we already have some information on the temporal trends in past infection hazard (though not its absolute magnitude) from estimates of the numbers of BSE-infected animals that were slaughtered for human consumption, but we are principally interested in predicting the future pattern of vCJD cases.

A further difference between the two analyses lies in the effect of additional exposure and/or host heterogeneities on infection risk and incubation period. Figure 10.1 presents a structural overview of the many factors determining the scale and pattern of the vCJD epidemic. For the sake of model simplicity, we collapse much of

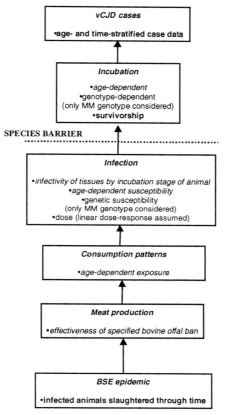

Figure 10.1 *Overview of factors influencing human risk of infection with vCJD. Available data are highlighted in bold and the parameters varied in the survival model in italics.*

the detail in the exposure phase of the flowchart into a few key (largely unknown) parameters. As more data become available to allow robust parametrization of, say, consumption patterns, the model structure itself could be extended, though the overall results of the scenario analysis would remain largely unchanged.

All TSEs exhibit complex relationships (Hunter *et al.*, 1989; Anderson *et al.*, 1996) between dose size, γ, infection risk and incubation period (here defined to be the time between infection and death) distribution, $f(u)$, and these are further modified by host characteristics such as PrP genotype, k, (see, for example, Goldmann *et al.*, 1991; Hunter *et al.*, 1989) and (potentially) age at infection, a'. We therefore allow the risk of infection to vary with these variables, defining $I_k(t, a', \gamma)$ to be the infection hazard at time t for a genotype k individual of age a' consuming dose γ. Convoluting this hazard with the corresponding IPD and integrating over all host (age at death, genotype) and infection (dose, time of infection) characteristics (while allowing for natural mortality) then gives the expected incidence of vCJD cases at time u:

$$
\begin{aligned}
c(u) \; = \; & \sum_k \int_0^{a_{max}} B_k(u-a) S(u, a) \\
& \times \int_{u-a}^u \int f_k(u-t, a-u+t, \gamma) I_k(t, a-u+t, \gamma) \\
& \times \exp\left[-\int_0^t \int I_k(t', a-u+t', \gamma')' d\gamma' dt' \right] d\gamma \, dt \, da
\end{aligned}
$$

where $B_k(t_b)$ is the birth rate of individuals of genotype k at time t_b and $f_k(u, a', \gamma)$ is the incubation PDF for an individual with genotype k infected at age a' with a dose γ.

The magnitude (though not the time-course) of the vCJD epidemic is, of course, primarily determined by the infection hazard, $I_k(t, a', \gamma)$. We simplify the calculation of this hazard by restricting our analysis to considering only the 40.1% (Owen *et al.*, 1990; Collinge *et al.*, 1991) of the British population that are methionine homozygous at codon 129 (MM_{129}) of the PrP gene, and by only considering infections caused by consumption (whether through food or other bovine products) of infected bovine tissue that occurred soon after animal slaughter (i.e. we do not consider human–human transmission or infection from other species).

The former assumption is justified by the fact that all vCJD cases to date have had genotype MM_{129} (suggesting increased susceptibility and/or shorter incubation periods for this genotype (Goldmann *et al.*, 1994; Raymond *et al.*, 1997; Zeidler *et al.*, 1997; Deslys *et al.*, 1998)), so that we have no data on which to model infection risk for other genotypes. If other genotypes have been infected, then our results give a lower bound on the maximum number of vCJD cases.

The latter assumption means that the infection hazard is determined by the pattern of human exposure to infected cattle tissue, and thus reflects the temporal pattern of the BSE epidemic and the effectiveness of the SBO ban. More precisely, the pattern of exposure depends on the numbers of infected animals slaughtered, the estimation of which has been outlined in Chapters 4 and 5 as well as in Anderson *et al.* (1996) and Ferguson *et al.* (1997a). These figures need to be stratified by bovine incubation stage, as infectivity is thought to vary over the course of the BSE incubation period (Fraser *et al.*, 1992; Wells *et al.*, 1994, 1998; Ministry of Agriculture, Fisheries and Food, 1996b; Spongiform Encephalopathy Advisory Committee, 1997). Figure 10.2(a) shows these estimates by quarter-year of date of slaughter and time to disease onset, while Figure 10.2(b) shows the same estimates weighted by an (example) exponentially increasing infectivity function that peaks at the end of the incubation period and gives a mean duration of infectiousness of 6 months. The difference is striking — while Figure 10.2(a) peaks in late 1988, Figure 10.2(b) peaks in 1992, after the introduction of the SBO ban. Hence, while the majority of infected animals were slaughtered before controls were put into place, slaughter rates of late incubation stage animals tracked the BSE case epidemic and hence peaked several years later.

For an MM_{129} genotype individual being exposed to dose γ at age a', the infection hazard at time t is

$$I(t, a', \gamma) = g_e(a')\nu(t)\beta(\gamma, a') \int h(\gamma|z, a')w(z, t)dz$$

where $h(\gamma|z, a')$ represents the probability density that an individual of age a' is exposed to a dose γ during consumption of tissue from an infected bovine slaughtered time z prior to disease onset; $w(z, t)$ is the proportion of cattle slaughtered at time t that are infected and time z away from disease onset; $g_e(a')$ is the mean frequency of beef consumption in people of age a'; $\nu(t)$ represents the effect of control measures at time t in preventing infectious material reaching the human food supply (here assumed to equally affect the probability of exposure to any dose); and $\beta(\gamma, a')$ is the probability that a dose of size γ will infect a human of age a'.

This complex dose-dependent model can be dramatically simplified if we make a number of justifiable assumptions. First, we assume the dose distribution has one of two forms:

- $h(\gamma|z, a') = h[\gamma/\Gamma(z, a')]/\Gamma(z, a')$, where $\Gamma(z, a')$ is a scaling factor that increases with z.
- $h(\gamma|z, a') = h[\gamma - \Gamma(z, a)]$, where $\Gamma(z, a')$ is an offset that increases with z, and $h(x) = 0$ for $x < 0$.

(a)

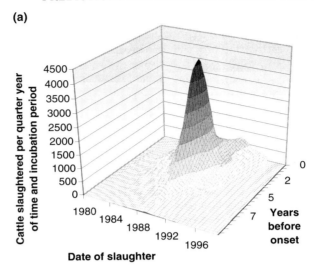

(b)

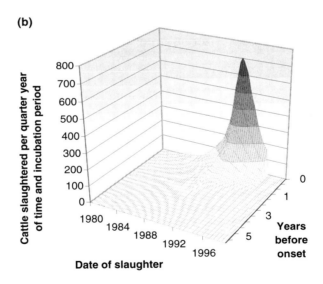

Figure 10.2 *(a) Number of BSE-infected cattle slaughtered, stratified by quarter year of slaughter and time remaining from slaughter date to when an animal would have exhibited clinical signs of BSE, measured in quarter years, as estimated from back-calculation models (see Chapters 4 and 5). (b) as (a), but weighted by relative infectivity, under the assumption that infectivity increases exponentially throughout the incubation period at a rate of 2/year, reaching a value of 1 at disease onset. Note this distribution peaks in 1992–93, after the introduction of measures designed to protect human health.*

For either form, Γ represents the effect of disease pathogenesis in cattle or age-dependent meat product consumption patterns on the dose distribution. The probability of exposure to larger doses increases with Γ.

Second, we assume that $\Gamma(z, a')$ is separable into time and age varying components:

$$\Gamma(z, a') = \Omega(z)g_r(a')$$

where $\Omega(z)$ is the relative infectiousness of an animal time z prior to disease onset, and $g_r(a')$ represents the potentially higher risk of age groups that consume more high-titre meat products.

The third simplifying assumption is of linear dose-response (supported by existing data (Diringer et al., 1998)). Thus $\beta(\gamma, a') = \beta\gamma g_s(a')$, where $g_s(a')$ represents the relative susceptibility of an individual of age a'.

Finally, we assume that the incubation period has an intrinsic variance much larger than any variability caused by dose (i.e. the incubation period distribution is effectively dose independent). We can then integrate out γ and obtain

$$I(t, a') = \nu(t)\beta g(a') \int \Omega(z)w(z, t)dz$$

where we have combined all age-dependency into a single normalized factor $g(a') = g_s(a')g_r(a')g_e(a')$ and have redefined β as a transmission coefficient (i.e. incorporating frequency of exposure).

The weakest assumption in this chain of reasoning is the last, as it is known from oral transmission studies of BSE in cattle (Anderson et al., 1996) that the incubation period distribution is strongly dependent on dose. The data are consistent with the hypothesis that mean incubation period decreases as the log of initial dose ($u \propto u_0 - r\log\gamma$), as would be predicted from a model of exponential growth of the aetiological agent in the host, with clinical onset (or death) occurring when some maximum threshold density is reached. However, it is likely that real-world doses experienced by cattle or humans did not span the large range (1–300 g) explored experimentally, but were probably at the lower end of the dose size spectrum – where a 1 or 2 log variability in dose would make a substantial difference to infection risk (due to linear dose-response), but have much less effect on the mean incubation period. This, coupled with the fact that incubation periods for TSEs are much more variable for oral exposure with a fixed dose than for intracerebral inoculation, means that assuming the incubation period distribution is independent of dose is not as unreasonable as might first be thought. Furthermore, by allowing the incubation period distribution to vary as a function of the age at infection, we are effectively approximating much more complex models that allow for age-dependence in the dose distribution (*e.g.* if young people were more likely to consume low-grade meat products — such as hamburgers or meat pies — that contained high titres of infectious tissue) and its corresponding effect on the incubation period distribution.

A further motivation for using the simplified expression for the infection hazard is its computational simplicity, together with the small volume of incidence data available with which more complex models could be tested. However, as more biological or epidemiological data become available, the more complex framework may prove useful in testing specific hypotheses regarding the detailed pattern of human exposure to infectious material.

Selection of potential functional forms for the incubation period distribution of vCJD was informed by the distributions observed for other TSEs, as discussed in Chapter 1. A particularly important feature that must be allowed for is the fact that several TSE incubation period distributions (including that of BSE) have relatively large minimum incubation periods. We therefore include the possibility of a substantial post-exposure delay prior to the onset of any vCJD cases. To explore the space of incubation period distributions we used a modified form of the 4 parameter generalized lambda distribution (Ramberg *et al.*, 1979). The generalized lambda distribution is described by the inverse CDF:

$$u(p) = \lambda_1 + \frac{p^{\lambda_3} - (1-p)^{\lambda_4}}{\lambda_2} \qquad (10.1)$$

where the λ_i are positive definite parameters. The generalized lambda distribution is useful because of its flexibility, enabling it to encompass virtually all the potential shapes of other more limited forms, such as offset Weibull, gamma and lognormal distributions (in terms of the range of third and fourth moment values spanned). However, the above distribution is only defined on $\lambda_1 - 1/\lambda_2 \leq u < \lambda_1 + 1/\lambda_2$, while we require a distribution that allows for incubation periods that are unbounded above. A suitable modified form is therefore

$$u(p) = \lambda_1 + \frac{p^{\lambda_3} + (1-p)^{-\lambda_4} - 1}{\lambda_2} \qquad (10.2)$$

which is defined on $\lambda_1 \leq u < \infty$. Note that while distributions described by inverse CDFs are difficult to treat analytically, they are relatively simple to evaluate numerically.

In the results presented here, bovine infectivity, $\Omega(z)$, is assumed to rise exponentially (from some baseline level) to reach a maximum at the end of the BSE incubation period — a trend consistent with currently available data on BSE and TSE pathogenesis (Fraser *et al.*, 1992; Wells *et al.*, 1994, 1998; Ministry of Agriculture, Fisheries and Food, 1996b; Spongiform Encephalopathy Advisory Committee, 1997). By varying two parameters (baseline level and exponential rate of increase) a wide variety of specific infectivity assumptions can be explored. The infection hazard is also strongly affected by the SBO ban introduced in November 1989. Our analyses assume that the ban was anywhere between 0 and 100% effective (with $\nu(t) = 1$ prior to this date).

As mentioned above, the analysis allowed for two forms of age-dependency — in susceptibility/exposure, and in the incubation period distribution. For the former, a variety of functional forms (normal, logistic, step) were explored, modelling the situation of exposure being largely limited to a particular age group. For the latter, we assumed that the incubation period increased with age using the scaling $f(u, a) = h(u)/s(a)$ where $h(u)$ is an age-independent incubation period density function and $s(a)$ is a scaling function. For simplicity, we use a logistic scaling function for the results presented below, namely

$$s(a) = \frac{\alpha_2 \exp(-\alpha_3 a) + \alpha_1}{\exp(-\alpha_3 a) + \alpha_1}$$

where $\alpha_1 > 0$, $0 \leq \alpha_2 \leq 1$ and $\alpha_3 > 0$. Epidemiologically, this form models the situation of young people having greater exposure to higher-titre products, being more likely to be infected with a higher dose and thus having a shorter incubation period.

10.3 Estimation/scenario analysis

There are two basic approaches that can be taken to exploring the space of possible epidemic scenarios consistent with the current incidence data:

- Maximum-likelihood parameter estimation, evaluation of model goodness of fit, and calculation of the 95% confidence region around the best-fit point.

- Treatment of each parameter combination as a separate 'model' (with no estimated parameters), and calculation of all 'models' consistent with the incidence data, as judged by comparing the likelihood ratio statistic with a χ^2 distribution with as many degrees of freedom as there are in the data.

The former approach* is most suited to applications where the number of independent data points is considerably larger than the number of parameters being estimated. When the number of parameters is close to or even exceeds the number that can be estimated from the data, formal parameter estimation becomes less useful as evaluation of goodness-of-fit statistics may not be possible, and in any case the 95% confidence region is likely to be so large as to give the best fit point relatively little relevance. In this situation − which is the one faced here − it can be preferable to adopt the latter approach, and just to sample (or, if possible, exactly determine) the space of parameter values (scenarios) consistent with the observed data. This has the benefit of avoiding potentially misleading point estimates being presented in contexts (*e.g.* media debates and political briefing documents!) where 95% confidence intervals are often forgotten. Furthermore, if the number of model parameters equals or exceeds the number of data degrees of freedom, this method imposes a stricter criterion for acceptance of parameter scenarios than that imposed by calculation of the 95% confidence region (the sets of accepted parameter points arising from the two approaches coincide only if the number of parameters equals the number of data points and the model fits the data exactly). The second approach is therefore the one adopted here, with a scenario (parameter combination) being accepted if both

---

* It should be noted that other work (Thomas and Newby, 1999) does attempt to use this method, but the conclusions drawn have little validity due to fundamental statistical and numerical errors (Ferguson *et al.*, 1999a) in the evaluation of the 'confidence limits' presented.

the age- and time-structured marginal case distributions are consistent at the 95% level with the corresponding observed marginal distributions as judged by the distribution of (Poisson) likelihood deviances obtained from parametric bootstrap sampling of the observed incidence data. We sample over a wide range of incubation period distribution shapes, age-dependent susceptibility/exposure functions $(g(a))$, SBO ban effectiveness levels $(\nu(t))$ and patterns of infectivity of cattle in different incubation stages $(\Omega(z))$, using Latin-hypercube sampling as an efficient method for exploring parameter space (Stein, 1987; McKay et al., 1979).

The range of possible outcomes sampled can then be illustrated using scatter plots that plot pairs of parameter values (or outcome statistics) for all accepted parameter combinations. However, it is important to note that such plots do not represent Bayesian posterior densities: i.e. the density of points in a particular region of parameter space should not be interpreted as representing the likelihood that the true parameters take these values. In the plots presented, we have frequently chosen the Latin-hypercube sampling weightings so as to examine a particular region of parameter space in more detail, or non-linearly transformed model parameters so as to optimize the efficiency of the sampling algorithm.

10.4 Determinants of epidemic size

The key factors determining the scale of the vCJD epidemic (at least in terms of numbers infected) are simply the numbers of infectious animals slaughtered for human consumption during the BSE epidemic, the degree of exposure and infectivity of bovine tissue to humans (characterized by β), and the size of the human population, N. However, for a given value of N and cattle infectivity profile, the incidence data act to put bounds on both the form of the incubation period distribution $f(u)$ and on the value of β. If the size of the susceptible population N is reduced, higher values of β are required to reproduce the same epidemic size.

This relationship is expressed here using the single parameter $r = \beta N/A$, the average number of individuals that will become infected from one maximally infectious (defined to be within 3 months of disease onset) animal, where A is the total rate of cattle slaughter. The value of r is the principal determinant of epidemic size, encompassing the relative infectiousness of different bovine tissues (Fraser et al., 1992; Wells et al., 1994, 1998; Ministry of

Agriculture, Fisheries and Food, 1996b; Spongiform Encephalopathy Advisory Committee, 1997), the infectivity to humans of these tissues (the species barrier) and the average number of susceptible individuals who will consume one carcass. While these parameters are largely unknown, given detailed data on the production, distribution and consumption of beef products and on the relative infectivity of different tissues, one could estimate an upper bound for r by estimating the maximum number of individuals who might have consumed tissue from a single carcass − since, irrespective of the absolute infectiousness of such tissue, this number would also be the maximum number of people who could be infected. Figure 10.3 illustrates the relationship between r and epidemic size for a range of assumptions about the rate at which bovine infectiousness increases during the BSE incubation period. Note that every point on this and later scatter plots represents a parameter set consistent with the incidence data observed to date.

Given a value for r, the IPD influences only the time-course, and not the size, of any epidemic. Demanding consistency with the incidence data then restricts the possible length and shape of the IPD. Conversely, if r is unknown but the incubation period is known, the incidence data can be used to restrict the possible values of r, and hence narrow the prediction intervals on total epidemic size. Both the length and the shape of the left-hand side of the distribution are important determinants. Small epidemics are generated from tight distributions with short modes, with the current incidence data representing a large proportion of the total distribution. Conversely, large epidemics are generated by long strongly peaked IPDs, for which the cases seen to date represent a small part of the left-hand tail of the distribution. Figure 10.4 illustrates this, plotting epidemic size against the mode of the IPD for a range of assumptions about the 'width' of the distribution. Note that we characterize the IPD by the mode, u_m, rather than the mean, as the mean can only be estimated once the vCJD epidemic is over (and the entire IPD has been seen), while estimation of the mode only requires the epidemic peak to have been reached. For the same reason, the width of the IPD is described by the 10th percentile up to the mode, $u_{m_{10\%}}$, defined by $F(u_{m_{10\%}}) = F(u_m)/10$ where $F(u)$ is the cumulative IPD, rather than the variance. Note that this quantity is solely a measure of the shape of the left-hand side of the IPD.

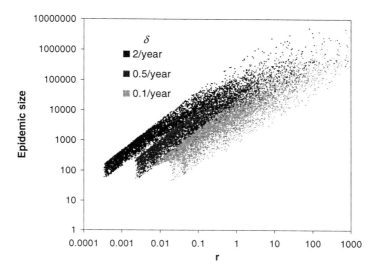

Figure 10.3 *Scatter plots of sampled scenarios showing the total epidemic size against r, the mean number of humans infected by one maximally infectious bovine, for three values of the exponential rate of growth of infectiousness in the bovine incubation period $(\Omega(z) = \exp(-\delta z)$, where z is the time to onset): $\delta = 2$, 0.5, and 0.1/year. Results are shown for the age-dependent susceptibility/exposure model (similar results are obtained assuming an age-dependent incubation period). As the infectiousness growth rate, δ, increases, the value of r required to produce a given epidemic size increases, since a decreasing fraction of cattle survive long enough to reach close to maximal infectiousness. This effect becomes less significant for mean durations greater than about 2 years due to cattle survivorship patterns (most cattle are slaughtered just after 2 years of age). Note that the figures do not represent posterior densities.*

10.5 Age structure of the epidemic

The strong clustering of vCJD cases in the 20–40-year-old age group means that epidemic scenarios consistent with the incidence data cannot be obtained without assuming age-dependency in either the incubation period or in susceptibility/exposure. More specifically, this is because the probability of observing no cases in patients over 53 years by chance is very small. Older individuals therefore need to have a longer incubation period and/or reduced susceptibility or exposure to the BSE agent.

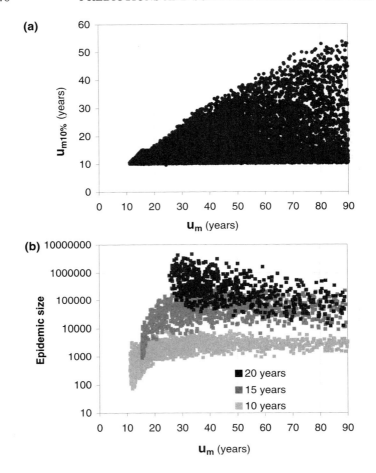

Figure 10.4 *Scatter plot of sampled scenarios showing (a) IPD mode,* u_m, *against the 'width' of the left-hand side of the IPD, as charac-terized by* $u_{m_{10\%}}$ *(see text) (b) IPD mode,* u_m, *against epidemic size for* $u_{m_{10\%}} = 10$ *years,* $u_{m_{10\%}} = 15$ *years and* $u_{m_{10\%}} = 20$ *years. Results are only shown for the age-dependent susceptibility/exposure model/indexAge-dependent susceptibility/exposure (similar patterns are observed for the age-dependent incubation period model).*

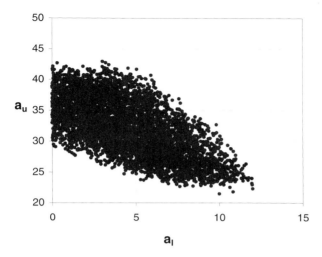

Figure 10.5 *Scatter plot of sampled scenarios showing lower age bound, a_l for susceptibility/exposure to vCJD infection against upper age bound, a_u, for the age-dependent susceptibility/exposure model discussed in the text.*

Assuming a uniform age-dependent susceptibility/exposure function between ages a_l and a_u, scenario analysis indicates (Figure 10.5) that a_u needs to be at least 21 years (to get sufficient cases in the $20-40$ age group), but less than 43 years (to get few enough cases over 50 years of age).

Assuming an age-dependent IPD but no age dependence in susceptibility/exposure, a similar analysis indicates that any significant increase in incubation period with age-at-infection must have been restricted to those aged over approximately 25 years at their time of infection.

10.6 Minimum incubation period

In the absence of significant age-dependent susceptibility/exposure, the minimum observed age of death of 19 years can only be explained by a comparable minimum incubation period for vCJD.

Even if age-dependent susceptibility/exposure is allowed for, the scenario analysis suggests (Figure 10.4) a lower bound on the *modal* incubation period of approximately 10 years. This is explained by comparing the temporal patterns of human exposure (namely the numbers of infected animals slaughtered for human consumption through time) with those of vCJD cases − which indicates that the time at which the first cases of vCJD were seen (1995) must to some extent be correlated with the time at which infection risk in cattle products first became significant − around the time the BSE epidemic began (some time between 1980 and around 1982−3). Put another way, the fact that no vCJD deaths prior to 1995 have been reported means that a large proportion of cases seen to date must have arisen from exposure in the first few years of the BSE epidemic − of course assuming that no unreported vCJD cases occurred prior to 1995. For the scenarios including age-dependent incubation periods, the fraction of cases reported through 1997 that were infected prior to 1986 never fell below 0.5, while for age-dependent susceptibility/exposure the lower bound on this fraction is 0.2. Interestingly, these values reconcile press reports that one vCJD case had been a vegetarian since 1986 with the basic hypothesis that vCJD was caused by exposure to infectious bovine products.

Further insight into the existence of a lower bound on the incubation period can be demonstrated through approximation of the full survival model with a simplified, analytically tractable form. We first assume that only cattle in the last 6 months of BSE incubation are infectious and that the SBO ban was completely ineffective. These assumptions are conservative in that they give the lowest incubation period bounds, but have the advantage of allowing the relative infection hazard for vCJD (the number of late incubation stage animals slaughtered through time) to be closely approximated by a Lorenz distribution (back-to-back exponentials) with risk beginning in 1980 and peaking in 1992. The ratio of vCJD-induced deaths occurring in 1996−98 to those occurring prior to 1996 can then be approximated by

$$\frac{\int_{1980}^{1998} F(1998 - t)exp(\alpha|1992 - t|)dt}{\int_{1980}^{1995} F(1995 - t)\exp(\alpha|1992 - t|)dt} - 1 > R_L$$

where $R_L = 9.78$ is the lower 95% confidence bound on the ratio obtained from the incidence data, $F(u)$ is the cumulative IPD function and $\alpha = 0.55$. In the (unrealistic) case of a fixed incubation period, this equation can be solved to give a lower bound of 6.1 years.

10.7 Epidemic predictability

With the constraint of 39 cases by 1 January 1999 epidemic sizes from consistent parameter scenarios range from 39 to around 10 million cases, clearly demonstrating that the current time series of vCJD cases contains too little information to allow useful predictions to be made at present. At the time this analysis was first performed (early 1998), similar results highlighted that speculation that a yearly time series of reported cases running 3 (in 1995), 10 (in 1996) and 10 (in 1997) was indicative of a small overall epidemic was totally unfounded. Such speculation has now lessened, perhaps because 16 vCJD deaths occurred in 1998, though these additional data do not help to reduce prediction uncertainty significantly. It proved useful to demonstrate this point to non-statisticians by stochastically generating epidemic scenarios from accepted parameter combinations that exactly reproduced yearly time series of cases to date (Figure 10.6). While not strictly informative in a statistical sense, such examples put the limited portion of the epidemic seen so far into perspective.

Given that current information is too limited to say anything sensible about future total epidemic size, it is interesting to ask how the pattern of cases observed over the next few years will aid prediction. Our scenario analyses give prediction intervals of $7-82$ cases in 1999 and $0-138$ cases in 2000. However, the upper bound on the total epidemic size is only substantially reduced if fewer than around 30 deaths from vCJD occur in this period. Above this value, the range of possible epidemic sizes is large.

Table 10.1 examines to what extent the numbers of vCJD cases reported over the next 2 years will aid prediction, and demonstrates − consistent with the discussion of epidemic size determinants early in this chapter − that predictability is only substantially enhanced if additional information enabling r (the infectiousness of bovine tissue to humans) to be bounded becomes available. A maximum of 50 cases observed in 1999 constrains future epidemic size to be small (less than 5000 cases) only if r is less than around 0.01. For r greater than this, large epidemics are still possible, even if fewer than 15 cases are confirmed in 1999. Only by the year 2000 does predictability improve, and only then if a small number of cases are observed in the period $1999-2000$. In that case, if a maximum of 29 cases were seen in 1999 and 2000, the total epidemic size is predicted to be less than 15,000 cases.

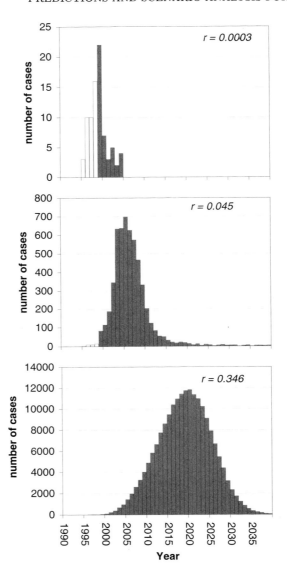

Figure 10.6 *Time series for three simulated epidemic scenarios. The stochastic realizations selected (30 were generated for each accepted scenario) all exactly match the incidence data time series to date (shown with white bars).*

Table 10.1 *Upper and lower bounds on total vCJD epidemic size strati-fied by the number of cases reported in 1999, 1999–2000 and r, the mean number of people infected by one maximally infectious bovine. Scenarios were accepted if the deterministic mean numbers of cases were consis-tent with current incidence data. The results are based on 30 stochastic realizations of each accepted parameter point. Bovine tissue infectivity was assumed to rise exponentially throughout the BSE incubation period, at a rate $\delta = 2/year$. The ranges were obtained with SBO ban effective-ness levels between 0 and 100%. If the SBO ban is assumed to have been over 90% effective, upper bounds are reduced by 5–10 fold for all but the largest band of r values.*

r	Cases 1999 4–14	15–29	30–49
0–0.01	50–3400	54–4500	69–4700
0.01–0.1	660–30,000	640–33,000	750–42,000
0.1–1.0	3800–180,000	3900–340,000	4100–340,000
1–10	25,000–320,000	$24{,}000-1.5 \times 10^6$	$25{,}000-1.6 \times 10^6$
10–100	220,000–760,000	$98{,}000-5.1 \times 10^6$	$100{,}000-6.1 \times 10^6$
100+	–	$330{,}000-9.0 \times 10^6$	$340{,}000-9.6 \times 10^6$

r	Cases 1999–2000 10–29	30–49	50–69
0–0.01	50–1600	69–3600	89–4000
0.01–0.1	660–6000	640–28,000	760–31,000
0.1–1.0	3900–14,000	3800–19,000	3900–340,000
1–10	–	24,000–420,000	$25{,}000-1.5 \times 10^6$
10–100	–	98,000–750,000	$97{,}000-2.0 \times 10^6$
100+	–	–	$340{,}000-2.1 \times 10^6$

10.8 Application to UAT programme design

The recent development of diagnostic tests able to identify abnor-mal prion protein in biopsies taken from vCJD patients prior to the onset of symptoms (Schreuder *et al.*, 1996; Hill *et al.*, 1997b) offers the potential of UAT of tonsillar and appendix tissues to estimate the infection prevalence in the population (Hilton *et al.*, 1998). We now discuss how the scenario analyses described above can inform the   design   of   the   testing   programme   and   the   test

sensitivity required to allow such programmes to provide meaningful estimates of overall infection prevalence in the British population.

10.8.1 Sample size and test sensitivity

The sensitivity of biopsy tests for PrP^{Sc} remains uncertain, and
in particular whether they are able to detect infectivity from an
early enough stage in the vCJD incubation period to allow reasonable estimates of overall — rather than late-stage — prevalence
to be made. If such tests are only able to identify late-stage infection, then their potential for reducing the current uncertainty in
future epidemic size will clearly be limited. As an example, let us
assume that a test can only diagnose infection in the last 3 years of
the incubation period. The scenario analyses presented above give
estimates of such late-stage prevalence to be between 0.4 and 24
infections per million people in 1998. It is therefore a straightforward exercise to calculate that to be 90% certain of detecting one
or more infections assuming the highest prevalence in that range
would require approximately 96,000 individuals to be tested, while
to detect the lowest prevalence would require testing a large proportion of the population. Furthermore, Figure 10.7(b) shows that
in the absence of information on r (the infectiousness of bovine tissue to humans), the estimated prevalence of infected individuals in
the second half of their incubation period in 1998 does not greatly
improve the predictability of the epidemic as a whole.

If tests are able to detect infection early in the incubation period, as suggested by experimental scrapie models (Kimberlin and
Walker, 1988), then Figure 10.7(a) shows that predictability does
improve, but that very large-scale testing is still required (Ghani *et
al.*, 1998b). However, so long as the duration of the incubation period during which infectivity is detectable remains unknown, this
straightforward analysis demonstrates that the possibility of a large
future epidemic cannot be ruled out, irrespective of the results of
any UAT programme, since the range of epidemic sizes consistent
with any small measured prevalence will remain large.

Given this uncertainty regarding sensitivity, it is reasonable to
ask what purpose would be served by undertaking a large-scale
UAT programme. The benefits largely lie in the possibility that
a significant prevalence might be measured. In that situation, it
would be reasonable to assume sensitivity was high, but more

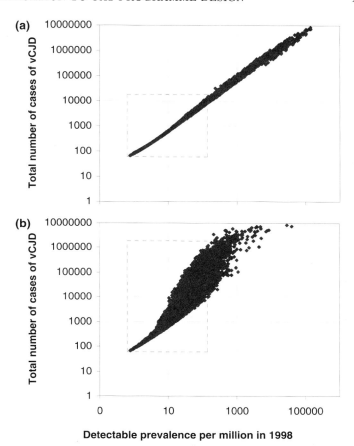

Figure 10.7 *Tests able to detect infection in (a) the last 75% and (b) the last 50% of the incubation period. Each point represents a simulated scenario that is consistent with the incidence data, and the confidence intervals on epidemic size if one infection is detected in a sample of 40,000 are shown.*

importantly, it would potentially give policy makers additional time to plan health care provision and direct increased funding toward research and development of drug treatments aimed at halting or curing disease pathogenesis, though the specificity of the test to clinical disease is not known.

10.8.2 Age-targeted UAT

Given the potential scale of the vCJD epidemic, it was decided in 1999 that the additional information that might be gained from UAT programmes outweighed the possible limitations of such surveys. It is therefore important to design the testing programme to optimize the informative value of the data collected. The role of the analyses presented in this chapter come from the observation that, in the absence of other prior information, the design of the most effective prevalence study should be based on the examination of the wide range of future vCJD scenarios that are consistent with:

- the time-series of the 39 vCJD cases seen to the end of 1998 (0, 3, 10, 10, 16)

- their age distribution

- the epidemiological profile of the BSE epidemic in cattle

Analysis of these scenarios requires consideration of the age-specific prevalences rather than simply the prevalence of infection in the population as a whole. For any particular study population chosen, the infection prevalence in a random sample of tonsillar or appendix tissues will be a weighted combination of the age-specific prevalences. Given that funding is only available to test a subset of all biopsy samples available, the most informative study population is the age group, or combination of age groups, in which the prevalence is most correlated with total epidemic size. Furthermore, the stored samples available were removed over a period of years, so it is important to test samples removed at the time when the correlation between measured prevalence and overall epidemic size is greatest.

Linear regression (on log-transformed data) is a simple approach to identify the age group in which prevalence is most correlated with epidemic size based on the R^2 statistic, thus minimizing the variance of prediction. The analysis is repeated over a range of sample collection times. Scenarios with age dependency in susceptibility/exposure or in the IPD need to be considered, as the two possibilities cannot yet be distinguished.

For each epidemic scenario, the detectable prevalence of infection was calculated in five-year age groups (0–4, 5–9, 10–14 years, etc) at different points in calendar time. Figure 10.8 presents the R^2 statistic over time assuming that the age distribution of the 39 vCJD cases to the end of 1998 is caused by age-dependent susceptibility/exposure, for three different age groups. For each scenario, the correlations between detectable prevalence and epidemic size are shown assuming that infection is detectable only in either the last 50% or 75% of the incubation period.

For age-dependent susceptibility/exposure, if the test is able to detect infection in the last 75% of the incubation period, the most informative age group to test is always tissues taken from 25–29-year-olds, although a random sample from 10 to 29 years performs nearly as well. If the test is only able to detect infection in the last 50% of the incubation period, then an older age group (30–34 or 35–39 years) is sometimes better. However, if any of these age groups are chosen, then the most informative tissues to test are those removed most recently.

Since tonsillar tissues will be mostly taken from children, it is relevant to examine the potential of a random sample of tissues taken from 10 to 14 years (tissues taken from younger children will be even less useful). The correlation between detectable prevalence in this age group and final epidemic size is far more variable than for older age groups, and hence testing tissues in this age group will be less informative. However, if samples from this age group have to be used then the most informative samples to test will be those removed earlier (*e.g.* between 1993–97), as shown in Figure 10.8. This arises because those children born from the late 1980s onward are not old enough both to have potentially been exposed to infection and to be in the later stages of incubation (when infection is assumed to be detectable).

Figure 10.9 presents the R^2 statistic over time assuming that the age distribution of the 39 vCJD cases is caused by an age-dependent incubation period. Under this scenario the most informative age group is generally slightly older (35–39 years) although there is very little difference between this age group and the 25–29 year group. Again, for this age group the most informative samples to test are those removed most recently. Testing in the 10 to 14 year group is, as in the age-dependent susceptibility/exposure

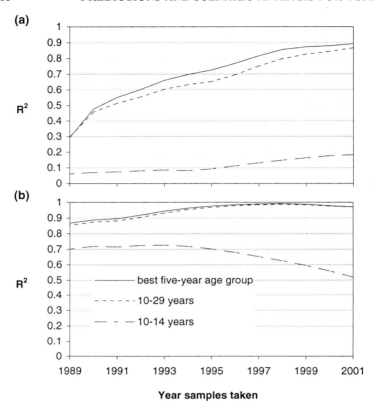

Figure 10.8 R^2 statistic (showing correlation between detectable prevalence in a sample of 1000 biopsies and total epidemic size) against time of sample collection for three age groups of sample individuals: 10–29 years, 10–14 years, and the most informative five-year age group. (a) if test is sensitive in the last 50% of the incubation period, or (b) if test is sensitive in the last 75% of the incubation period. Results are shown for the age-dependent susceptibility/exposure model.

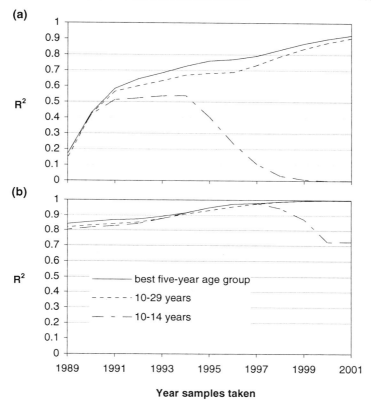

Figure 10.9 R^2 *statistic (showing correlation between detectable prevalence in a sample of 1000 biopsies and total epidemic size) against time of sample collection for three age groups of sample individuals: 10–29 years, 10–14 years, and the most informative five-year age group. (a) if test is sensitive in the last 50% of the incubation period, or (b) if test is sensitive in the last 75% of the incubation period. Results are shown for the age-dependent incubation period model.*

case, less informative than testing in older individuals close to the present time.

In both forms of age-dependent epidemic models, a random sample is only slightly less informative than the most informative age groups. As in age-targeted samples, a random sample of biopsy tissues is most informative on recently removed tissues.

Table 10.2 *Upper and lower bounds on total vCJD epidemic size consistent with detecting between 0 and 20 infections in a UAT random sample of patients in all age groups, assuming the test is able to detect infection in the last 75% of the incubation period (second column), or that test sensitivity is completely unknown (third column).*

Number of infections detected	Range of epidemic sizes assuming infection detectable in last 75% of incubation period	Conservative range of epidemic sizes
0	58–640,000	$58-8.6 \times 10^6$
1	1900–930,000	$1900-8.6 \times 10^6$
2	$19,000-1.1 \times 10^6$	$17,000-8.6 \times 10^6$
3	$52,000-1.3 \times 10^6$	$43,000-8.6 \times 10^6$
4	$86,000-1.6 \times 10^6$	$69,000-8.6 \times 10^6$
5	$140,000-1.8 \times 10^6$	$100,000-8.6 \times 10^6$
10	$300,000-2.9 \times 10^6$	$290,000-8.6 \times 10^6$
20	$840,000-4.2 \times 10^6$	$800,000-8.6 \times 10^6$

10.8.3 Information gained from UAT and the effect of sample size

To judge the informative value of any particular UAT programme, it is necessary to judge how a particular programme result restricts the range of consistent future epidemic scenarios. Let us assume that initially a sample of 1000 tissues is tested. As the model only considers the epidemic in the population that is MM_{129} (approximately 40% of the population), the effective population size is therefore 400. Consistency of model scenarios with the study result is then assessed by selecting scenarios that fall within the exact binomial 95% confidence bounds for the prevalence of detectable infection calculated for each possible study outcome (0, 1, 2 etc. infections detected).

Table 10.2 presents the range of possible total epidemic sizes if we test a sample of 1000 tonsillar tissues from all age groups that were removed in 1998, assuming infection is detectable in the last 75% of the incubation period.

In the absence of knowledge on test sensitivity, the most conservative approach is to assume that the test is able to detect

Table 10.3 *Upper and lower bounds on total vCJD epidemic size consistent with detecting between 0 and 20 infections in a UAT random sample of patients in all age groups, assuming test is able to detect infection in the last 75% of the incubation period, for sample sizes between 2000 and 20,000.*

Infections detected	Sample Size		
	2000	5000	20,000
0	58−320,000	58−140,000	58−64,000
1	990−540,000	530−190,000	263−90,000
2	9100−640,000	3600−250,000	1800−140,000
3	25,000−660,000	9100−320,000	4600−150,000
4	43,000−930,000	17,000−340,000	8200−180,000
5	64,000−930,000	25,000−410,000	12,000−190,000
10	$200,000−1.6 \times 10^6$	75,000−640,000	37,000−320,000
20	$380,000−2.5 \times 10^6$	200,000−970,000	94,000−540,000

infection throughout the incubation period to estimate a lower bound on epidemic size, and to assume that the test is only able to detect infection late in the incubation period to estimate an upper bound. This approach was used in calculating the third column of the above table.

Clearly, the power of any survey can be increased by increasing the sample size. Table 10.3 indicates the range of epidemic sizes consistent with different survey results for a range of potential sample sizes. Again, it is assumed that samples are drawn randomly from the tonsillar tissues removed in 1998, and that the test is able to detect infection in at least the last 75% of the incubation period.

If age groups with lower expected prevalence (for example 10−14 years) are tested and two or more infections are detected, then this will generally narrow the bounds on the range of epidemic sizes but indicate larger epidemics.

10.9 Conclusions

Those with a statistical background might reasonably wonder as to the purpose of performing extensive analysis of time-series data on only 39 cases of a new disease — it being 'intuitively obvious' that with such limited information it must be virtually impossible to predict the future extent of any epidemic where the incubation period might be many years. However, one reason for the above analysis being performed was that, despite what intuition might tell us, there was considerable speculation regarding the scale of the epidemic, both in the media and within the scientific and health communities. In the minds of some policy makers, no scientific or statistical conclusions should be treated as 'obvious' unless backed up with extensive scientific analysis. Hence, one key purpose of the work presented here was to demonstrate the extent of current uncertainty, and thus quash ill-informed speculation that the few cases seen to date necessarily implied that the whole vCJD epidemic would be small (however that word was defined). It should be emphasised that our analyses similarly do not prove the converse speculation — that, for what ever reason (*e.g.* 750,000 infected animals were consumed, so many people *must* have been infected), the vCJD epidemic is inevitably going to be 'large.'

With the recent trend toward the commissioning of complex risk assessments during any decision-making process affecting public safety, work demonstrating the many difficulties of such exercises in the case of human exposure to BSE is not always welcome. Most politicians are much more comfortable quoting point estimates than admitting to large underlying uncertainty, in part because of the difficulty in forming and justifying policy options when risks cannot be accurately quantified. This type of exhaustive sensitivity analysis (over 5 million parameter scenarios were sampled here) is therefore an extremely useful tool to highlight the great uncertainty about key determinants, and is essential if policy makers are to fully understand the limitations in current information about risks to the human population. The public debate and policy debacle relating to the possible risks to the population from dorsal route ganglia attached to meat cuts on the bone from BSE-infected cattle (Comer, 1997) gave a prime example of the dangers inherent in ignoring such uncertainty and assigning of specific values to risk.

As well as quantification of uncertainty, we have also demonstrated in this chapter how scenario analysis can have practical use in optimizing UAT programme design. However, we have also shown that caution must be exercised in interpreting results from such surveys, since it is not known at what stage in the incubation period of vCJD infectivity can be detected. For example, a low prevalence will indicate a small epidemic if the test is able to detect infection throughout the incubation period, or alternatively, could indicate a larger epidemic if the test is only able to detect late-stage infection.

The mathematical framework relating the BSE epidemic and the magnitude and duration of the vCJD epidemic can be extended to encompass further complications as new evidence emerges. The current model is based on two key assumptions — that the incidence of vCJD is correlated with the pattern of the BSE epidemic in cattle, and that the IPD of vCJD for the MM_{129} genotype is unimodal with a broadly similar form to other TSEs. However, given the observed properties of TSEs in animal models, it seems likely that host genetic background influences the incubation period, that susceptibility may be age-dependent and that the incubation period may depend both on age at exposure and dose of the agent. Once more data become available to estimate parameters relating to such effects, the survival model outlined above will be easily extensible to cope with such additional biological detail. Similarly, while we currently lack data on the distribution of exposure risk in the human population (in particular the distribution of animals and animal parts in meat products) and on how risk is related to the duration, magnitude and frequency of exposure to the agent, incorporation of such information (when available) offers the potential to refine the range of scenarios consistent with the incidence data. A further additional refinement would be to model the potential effect of vCJD within the human population, for example through transfusion/transplantation of material from infected donors.

However, none of this additional detail is likely to substantially change the basic conclusion that the observed time series of vCJD cases and their age distribution are consistent with a total epidemic size ranging from that already seen to a large proportion of the susceptible population. This is because the final epidemic size is primarily determined by bovine infectiousness to humans, characterized here by r, and (for obvious ethical reasons) this quantity is never going to be directly measured. That said, one potentially useful avenue for future research is to determine an upper bound on r, since we have shown that this is a critical determinant of future epidemic predictability. This could be achieved by combining data on how animals are distributed in the food manufacturing process with population-based surveys of dietary habits — in essence calculating how many people were likely to have consumed the potentially most infectious tissue from any one animal.

CHAPTER 11

Future directions

11.1 Updated back-calculation analysis of BSE in Great Britain

Decisions on the resumption of exportation of cattle from Great Britain will require the assessment of any remaining public and animal health risks to be addressed internationally. In particular, European Union member states will require reassurance that the BSE epidemic is indeed in permanent decline. There is therefore a continuing need to periodically update predictions of future case numbers on the basis of the new case data available.

It was announced in the House of Commons in December 1996 that a selective cull would be undertaken targeting animals from the same cohort as BSE cases, compulsory for such animals born between 16 October 1990 and 30 June 1993 and voluntarily for such animals born between 1 July 1989 and 15 October 1990. As a result, any updated analysis requires estimation of the selective cull's effectiveness at reducing case incidence.

We define $[1 - \Lambda_S(t|t_0)]$ to be the probability that a case onsetting at time t in an animal born at time t_0 is prevented by the selective cull. Thus, the PDF for an animal born at time t_0 being reported as a case at age a is given by

$$
\begin{aligned}
\phi_{RC}(t_0, a) \;=\;& \Lambda(t_0 + a)\Lambda_S(t_0 + a|t_0)S(a) \\
& \times \sum_g p_g \int_0^a \sum_{j=F,M,H} \rho_{jg}(a - u|t_0)f_j(u)du.
\end{aligned}
$$

We define $\Lambda_S(t|t_0)$ to equal a constant, denoted Λ_S, for those animals in the 1990, 1991, 1992 and 1993 cohorts after time t_S, and to equal zero otherwise. For simplicity, we did not differentiate between the compulsory and voluntary components of the culling scheme.

The back-calculation model was fitted to data available through mid-1998, with cases arising from the 1974−95 birth cohorts, with

Table 11.1 *The fitted and predicted incidence of cases of BSE for the years 1996 to 2001 in Great Britain based on data through mid-1996 and based on data through mid-1998 for the model with no maternal and no horizontal transmission (with 95% confidence intervals) compared with the observed incidence of cases available through January 1999.*

Year	Fits/Predictions using data through mid-1996	Fits/Predictions using data through mid-1998	Observed incidence
1996	8452 (7673,9355)	7424 (7172,7694)	7432
1997	5125 (4136,6381)	4649 (4386,4944)	4236
1998	2628 (1819,4011)	3088 (2638,3753)	2807
1999	1090 (659,2169)	1017 (603,1638)	
2000	380 (211,943)	391 (215,654)	
2001	118 (63,337)	124 (66,211)	

and without estimation of the effectiveness of the selective cull. Predictions of future incidence in Great Britain are presented in Table 11.1, assuming $\Lambda_S(t|t_0) = 1$ for all cohorts, an incubation period distribution of form C and an age-dependent susceptibility/exposure distribution of form 7. This fit had a likelihood ratio goodness-of-fit statistic of 699, compared with that of 551 for the model with $\Lambda_S(t|t_0) = \lambda$ for cohorts 1990−93 from January 1997 onward. Although the latter produced a much better fit to the analysed data, more recent incidence data (as yet incomplete and thus not analysed) indicates that the model allowing for an effect of the selective cull underestimates current incidence levels.

The key uncertainty is whether the demography, specifically age-specific survival probabilities, was affected by the over 30 months slaughter scheme whereby after March 1996 only animals under 30 months of age were slaughtered for consumption. Even very small changes in these survival probabilities have a big impact on the fit of the back-calculation models. Thus, additional data on recent demographic trends will be required for robust analysis of the most up-to-date BSE incidence data.

There are three key control measures that reduce any residual risk to the British public from the consumption of meat from BSE-infected cattle: the specified offal ban introduced in November 1989, the ban on the slaughter for consumption of animals over 30 months of age, indexban on the slaughter for consumption of animals over 30 months of age and the ban on the sale of beef on the bone (designed to prevent risk from the consumption of dorsal route ganglia) (Spongiform Encephalopathy Advisory Committee, 1997). Updated model fits are key to the evaluation of the reduction in risks due to each ban, particularly in the assessment of infection incidence among young cattle, since inevitably with back-calculation models the hazard of feed-borne transmission is most poorly estimated in recent years.

Back-calculation model estimates of infected animals slaughtered stratified by age will inform decisions on the eventual relaxation of the ban on the sale of beef on the bone, the ban on the consumption of meat from animals over 30 months of age and the ban on the exportation of British beef and live cattle. However, the remaining uncertainty in the recent demography will need to be addressed before robust estimates can be obtained from the most up-to-date incidence data.

11.2 vCJD epidemic prediction

As the BSE epidemic continues to decline, focus centers on the potential scale of the vCJD epidemic. Via vCJD, the poor political and scientific management of the BSE epidemic has left a legacy of increased public concern about food safety, and mistrust of government pronouncements on risk. In response, there is an increasing use of formal risk analysis procedures to inform future policy development (beef on the bone (Comer, 1997) and genetically modified food being prime examples). Since the quality of BSE-related risk analyses performed in the past has left much to be desired, it is incumbent on statisticians to ensure that the current and future analyses of the vCJD epidemic be as robust and thorough as possible, and do not under-emphasise the uncertainty around any prediction of future case numbers.

As can be seen from Table 10.1, uncertainty about the size of the vCJD epidemic is likely to remain for some years to come. Even if relatively few cases are seen in 1999 and 2000, the confirmation of vCJD in an individual of another genotype (all vCJD cases confirmed by April 1999 have been in individuals homozygous for methionine at the 129 codon of the prion protein gene) will suggest the possibility of concurrent epidemics within the British population.

As data from the unlinked anonymous testing (UAT) programme become available, the scenario analysis framework will be expanded to test parameter sets simultaneously for consistency with the incidence time series and the UAT results. If the UAT data are stratified by age, we will gain more insight into the process underlying the observed age structure of the vCJD cases.

Data collected in the next few years may yield insight into possible seasonality of deaths among vCJD cases and potential risk factors such as geographic region. It is unlikely though, even with relatively high incidence, that insight will ever be gained into dietary risk factors for the disease since reliable dietary data are notoriously difficult to collect even for relatively recent periods. The collection of meaningful data from relatives about the cases' previous dietary habits will be even more difficult with the second-hand nature of the observations and the potential for recall bias.

11.3 Scrapie epidemiology

The BSE and vCJD epidemics in Great Britain have increased concern in both the scientific community and the general public about scrapie. Although there is no evidence that scrapie has ever crossed the species barrier, one hypothesis for the origin of the BSE epidemic is that a strain of scrapie was transmitted from sheep to cattle. Recent work has focussed on the need to monitor and control scrapie with a future goal of eradication.

A further reason for concern about TSEs in sheep is that many sheep in Britain were fed potentially contaminated MBM prior to the introduction of the ruminant feed ban in 1988 and the final ban on the use of MBM in all animal feeds in 1996. Given the successful experimental transmission of BSE to sheep (Foster *et al.*, 1996) by oral challenge with infected bovine brain tissue, authorities are concerned about the possibility that a BSE-like spongiform encephalopathy may have established itself in the sheep population, presenting an additional risk of transmission to humans.

In January 1993, scrapie became a notifiable disease in the European Union. Since this time, approximately 350 cases have been reported annually in Great Britain, corresponding to an annual incidence of approximately two cases per 100,000 sheep. The degree of under-reporting is, however, believed to be high. A survey suggests that cases may have occurred on over 25% of sheep farms (Morgan et al., 1990). However, a compulsory slaughter with compensation scheme recently introduced in Britain may improve reporting rates further.

The application of transmission models (e.g. Stringer et al. (1998), Woolhouse et al. (1998), and Woolhouse et al. (1999)) with parameters along the lines of those presented in Chapter 4 to age-structured scrapie incidence data both nationally, using the surveillance data collected by the Ministry of Agriculture, Fisheries and Food, and from intensively studied flocks will yield insights into the factors governing the spread and persistence of the disease. Recent work on diagnostic tests for scrapie (Schreuder et al., 1996, 1998) will allow age-specific preclinical prevalence survey data to supplement case incidence data allowing increasingly realistic models to be fitted. However, the apparent increased importance of direct horizontal transmission within flocks and the importance of genetically variable susceptibility to scrapie will make the computational burden of scrapie models even greater than that of BSE.

11.4 Conclusion

Understanding the transmission dynamics of a novel infectious agent on the basis of limited data has always posed challenges for researchers interested in infectious disease epidemiology. It is therefore unfortunate that progress has perhaps been slowed in the past by a lack of communication between biostatistics and mathematical biology. This is in part due to the rather different tools used by the two disciplines — survival analysis and generalized linear models in one case and non-linear dynamical modelling in the other — but also because of the rather different motivations of researchers in the two fields — on one hand to characterize significant interactions, estimate key parameters and predict future trends, and on the other to understand how the ecological and evolutionary processes that govern host-parasite systems shape transmission dynamics at the population level. However, from being relatively rare 20 years ago, research groups with skills spanning both disciplines are now increasingly common. Indeed, the ability to employ tools from both traditions is rapidly becoming a vital requirement for effective research in infectious disease epidemiology.

This book is the product of such a multi-disciplinary research effort, and we hope we have shown the benefits to be gained (both in terms of qualitative insight and quantitative understanding) from combining modern statistical techniques with dynamical models. The complexities posed by transmissible spongiform encephalopathies (TSEs) (including multiple routes of transmission, long incubation periods, complex pathogenesis, clustering of infections) indeed necessitated an approach.

Our basic statistical tool has been maximum likelihood estimation using survival models, but with the form of hazard adopted being strongly informed by an understanding of basic transmission processes and mechanisms. Such an approach is not without costs, however − arising primarily from the analytical complexity of the resulting models and consequent requirement for high-performance computer systems with which to perform model analysis and estimation. It is therefore important not to forget the important role that much simpler models can play in giving qualitative insight into potential mechanisms underlying observed pattern − as demonstrated by the discussion of BSE case clustering in Chapter 8.

The main limitation of the analyses presented was being unable to directly fit a non-linear, stochastic model of BSE transmission dynamics to the case data. Back-calculation models, while powerful, remain linear models and largely ignore the combined effects of spatio-temporal correlation, non-linearity and stochasticity. The next few years seem certain to remove many of the computational and methodological barriers preventing such analyses at present, with Markov chain Monte Carlo (*e.g.* Gibson (1997) and Gibson and Renshaw (1998)) and moment closure approaches (*e.g.* Isham (1995) and Bolker and Pacala (1997)) appearing at present to be the most powerful routes to this goal.

Of course there are always some questions (here, for example, the origin of BSE and the scale of any future vCJD epidemic) that cannot be addressed from currently available data regardless of the manner of analysis, and in such cases we have tried to make explicit the limitations as well as the uses of the various modelling frameworks presented. We feel strongly that in some contexts — and particularly when informing policy making — it is important to quantify uncertainty and resist the temptation to 'take a best guess' thereby underplaying how little is known.

References

Alper, T., Cramp, W.A., Haig, D.A. and Clarke, M.C. (1967). Does the agent of scrapie replicate without nucleic acid? *Nature*, **214**, 764–766.

Alper, T., Haig, D.A. and Clarke, M.C. (1966). The exceptionally small size of the scrapie agent. *Biochem. Biophys. Res. Commun.*, **22**, 278–284.

Anderson, R.M., Cox, D.R. and Hillier, H.C. (1989). Epidemiological and statistical aspects of the AIDS epidemic: Introduction. *Philos. Trans. Roy. Soc. London, Ser. B*, **325**, 39–44.

Anderson, R.M., Donnelly, C.A., Ferguson, N.M., Woolhouse, M.E.J., Watt, C.J., Udy, H.J., MaWhinney, S., Dunstan, S.P., Southwood, T.R.E., Wilesmith, J.W., Ryan, J. B.M., Hoinville, L.J., Hillerton, J.E., Austin, A.R. and Wells, G.A.H. (1996). Transmission dynamics and epidemiology of BSE in British cattle. *Nature*, **382**, 779–788.

Anderson, R.M. and May, R.M. (1991). *Infectious Diseases of Humans: Dynamics and Control*. Oxford University Press, Oxford.

Bacchetti, P., Segal, M.R. and Jewell, N.P. (1993). Back calculation of HIV-infection rates. *Stat. Sci.*, **8**, 82–101.

Bailey, N.T.J. (1975). *The Mathematical Theory of Infectious Diseases and its Applications*. Griffin, London.

Baker, H.F., Ridley, R.M. and Wells, G.A.H. (1993). Experimental transmission of BSE and scrapie to the common marmoset. *Vet. Rec.*, **132**, 403–406.

Barhen, J., Protopopescu, V. and Reister, D. (1997). TRUST: A deterministic algorithm for global optimization. *Science*, **276**, 1094–1097.

Barlow, R.M. and Middleton, D.J. (1990). Dietary transmission of bovine spongiform encephalopathy to mice. *Vet. Rec.*, **126**, 111–112.

Bassett, H. and Sheridan, C. (1989). Case of BSE in the Irish Republic. *Vet. Rec.*, **124**, 151.

Berger, J.R., Weisman, E. and Weisman, B. (1997). Creutzfeldt-Jakob disease and eating squirrel brains. *Lancet*, **350**, 642.

Bolker, B. and Pacala, S.W. (1997). Using moment equations to understand stochastically driven spatial pattern formation in ecological systems. *Theor. Pop. Biol.*, **52**, 179–197.

Bovine Offal (Prohibition) Regulations (1989). *Statutory Instrument 1989 No. 2061*. HMSO, London.

Bovine Spongiform Encephalopathy Order (1988). *Statutory Instrument 1988 No. 1039.* HMSO, London.

Brookmeyer, R. and Gail, M.H. (1986). Minimum size of the acquired immunodeficiency syndrome (AIDS) epidemic in the United States. *Lancet,* **2**, 1320–1322.

Brookmeyer, R. and Gail, M.H. (1988). A method for obtaining short-term projections and lower bounds on the size of the AIDS epidemic. *J. Am. Statist. Assoc.,* **83**, 301–308.

Brown, P. and Bradley, R. (1998). 1755 and all that: a historical primer of transmissible spongiform encephalopathy. *Brit. Med. J.,* **317**, 1688–1692.

Brown, P. and Gajdusek, D.C. (1991). Survival of scrapie virus after 3 years' interment. *Lancet,* **337**, 269–270.

Brown, P., Goldfarb, L.G. and Gajdusek, D.C. (1991). The new biology of spongiform encephalopathy: infectious amyloidoses with a genetic twist. *Lancet,* **337**, 1019–1022.

Brown, P., Liberski, P.P., Wolff, A. and Gajdusek, D.C. (1990). Resistance of scrapie infectivity to steam autoclaving after formaldehyde fixation and limited survival after ashing at 360°C: practical and theoretical implications. *J. Infect. Dis.,* **161**, 467–472.

Brown, P., Preece, M.A. and Will, R.G. (1992). Friendly fire in medicine: hormones, homografts and Creutzfeldt-Jakob disease. *Lancet,* **340**, 24–27.

Bruce, M., Chree, A., McConnell, I., Foster, J., Pearson, G. and Fraser, H. (1994). Transmission of bovine spongiform encephalopathy and scrapie to mice: strain variation and the species barrier. *Philos. Trans. Roy. Soc. London, Ser. B,* **343**, 405–411.

Bruce, M.E. (1996). Strain typing studies of scrapie and BSE. In Baker, H.F. and Ridley, R.M., editors, *Methods in Molecular Medicine: Prion Diseases,* pp. 223–236. Humana Press, Totowa, New Jersey.

Bruce, M.E., McConnell, I., Fraser, H. and Dickinson, A.G. (1991). The disease characteristics of different strains of scrapie in Sinc congenic mouse lines: implications for the nature of the agent and host control of the pathogenesis. *J. Gen. Virol.,* **72**, 595–603.

Bruce, M.E., Will, R.G., Ironside, J.W., McConnell, I., Drummond, D., Suttie, A., McCardle, L., Chree, A., Hope, J., Birkett, C., Cousens, S., Fraser, H. and Bostock, C.J. (1997). Transmissions to mice indicate that 'new variant' CJD is caused by the BSE agent. *Nature,* **389**, 498–501.

BSE Inquiry (1999). *http://www.bse.org.uk/.*

Burger, D. and Gorham, J.R. (1977). Observation on the remarkable stability of transmissible mink encephalopathy virus. *Res. Vet. Sci.,* **22**, 131–132.

Burger, D. and Hartsough, G.R. (1965). Encephalopathy of mink. II.

Experimental and natural transmission. *J. Infect. Dis.*, **115**, 393–399.

Carolan, D.J.P., Wells, G.A.H. and Wilesmith, J.W. (1990). BSE in Oman. *Vet. Rec.*, **126**, 92.

Caughey, B. and Chesebro, B. (1997). Prion protein and the transmissible spongiform encephalopathies. *Trends in Cell Biol.*, **7**, 56–62.

Collinge, J., Harding, A.E., Owen, F., Poulter, M., Lofthouse, R., Boughey, A.M., Shah, T. and Crow, T.J. (1989). Diagnosis of Gerstmann-Straussler syndrome in familial dementia with prion protein gene analysis. *Lancet*, **2**, 15–17.

Collinge, J., Palmer, M.S. and Dryden, A.J. (1991). Genetic predisposition to iatrogenic Creutzfeldt-Jakob disease. *Lancet*, **337**, 1441–1442.

Collinge, J., Sidle, K. C.L., Meads, J., Ironside, J. and Hill, A.F. (1996). Molecular analysis of prion strain variation and the aetiology of 'new variant' CJD. *Nature*, **383**, 685–690.

Collins, S., Law, M.G., Fletcher, A., Boyd, A., Kaldor, J. and Masters, C.L. (1999). Surgical treatment and risk of sporadic Creutzfeldt-Jakob disease: a case-control study. *Lancet*, **353**, 693–697.

Comer, P.J. (1997). *Assessment of Risk from Possible BSE Infectivity in Dorsal Root Ganglia.* Det Norske Veritas (see http://www.dnv.com/).

Cousens, S.N., Linsell, L., Smith, P.G., Chandrakumar, M., Wilesmith, J.W., Knight, R. S.G., Zeidler, M., Stewart, G. and Will, R.G. (1999). Geographical distribution of variant CJD in the UK (excluding Northern Ireland). *Lancet*, **353**, 18–21.

Cox, D.R. and Oakes, D. (1984). *Analysis of Survival Data.* Chapman & Hall, London.

Cox, D.R. and Miller, H.D. (1965). *The Theory of Stochastic Processes.* Methuen, London.

Curnow, R.N. and Hau, C.M. (1996). The incidence of bovine spongiform encephalopathy in the progeny of affected sires and dams. *Vet. Rec.*, **138**, 407–408.

Curnow, R.N., Hodge, A. and Wilesmith, J.W. (1997). Analysis of the bovine spongiform encephalopathy maternal cohort study: The discordant case-control pairs. *Appl. Statist.*, **46**, 345–349.

Curnow, R.N., Wijeratne, W.V.S. and Hau, C.M. (1994). The inheritance of susceptibility to BSE. In Bradley, R. and Marchant, B., editors, *Proceedings of European Commission Consultation on Transmissible Spongiform Encephalopathies*, pp. 109–124. Brussels.

Davies, D.C. and Kimberlin, R.H. (1985). Selection of Swaledale sheep of reduced susceptibility to experimental scrapie. *Vet. Rec.*, **116**, 211–214.

Dawson, M., Wells, G.A.H. and Parker, B.N.J. (1990a). Preliminary evidence of the experimental transmissibility of bovine spongiform encephalopathy to cattle. *Vet. Rec.*, **126**, 112–113.

Dawson, M., Wells, G.A.H., Parker, B.N.J. and Scott, A.C. (1990b). Primary parenteral transmission of bovine spongiform encephalopathy to the pig. *Vet. Rec.*, **127**, 338.

De Jong, M. C.M., Diekmann, O. and Heesterbeek, H. (1995). How does transmission of infection depend on population size? In Mollison, D., editor, *Epidemic models: Their Structure and Relation to Data*, pp. 84–94. Cambridge University Press, Cambridge.

deSilva, R. (1996). Human spongiform encephalopathy: Clinical presentation and diagnostic tests. In Baker, H.F. and Ridley, R.M., editors, *Methods in Molecular Medicine: Prion Diseases*, pp. 15–33. Humana Press, Totowa, New Jersey.

Dealler, S.F. and Lacey, R.W. (1994a). Suspected vertical transmission of BSE. *Vet. Rec.*, **134**, 151.

Dealler, S.F. and Lacey, R.W. (1994b). Vertical transfer of prion disease. *Human Reproduction*, **9**, 1792–1796.

Denny, G.O. and Hueston, W.D. (1997). Epidemiology of bovine spongiform encephalopathy in Northern Ireland 1988 to 1995. *Vet. Rec.*, **140**, 302–306.

Denny, G.O., Wilesmith, J.W., Clements, R.A. and Hueston, W.D. (1992). Bovine spongiform encephalopathy in Northern Ireland: epidemiological observations 1988–1990. *Vet. Rec.*, **130**, 113–116.

Department of Agriculture and Fisheries for Scotland (1975–1980). *Agricultural Statistics – Scotland 1974–1979*. HMSO, Edinburgh.

Department of Agriculture and Fisheries for Scotland (1981–1990). *Economic Report on Scottish Agriculture 1980–1989*. HMSO, Edinburgh.

Deslys, J.-P., Jaegly, A., d'Aignaux, J.H., Mouthon, F., de Villemeur, T.B. and Dormont, D. (1998). Genotype at codon 129 and susceptibility to Creutzfeldt-Jakob disease. *Lancet*, **351**, 1251.

DiCiccio, T.J. and Efron, B. (1996). Bootstrap confidence intervals. *Statist. Sci.*, **11**, 189–228.

Dickinson, A.G., Meikle, V.M.H. and Fraser, H. (1968). Identification of a gene which controls the incubation period of some strains of scrapie in mice. *J. Comp. Pathol.*, **78**, 293–299.

Dickinson, A.G. and Outram, G.W. (1988). Genetic aspects of unconventional virus infections: the basis of the virino hypothesis. *Ciba Foundation Symp.*, **135**, 63–83.

Diringer, H., Roehmel, J. and Beekes, M. (1998). Effect of repeated oral infection of hamsters with scrapie. *J. Gen. Virol.*, **79**, 609–612.

Donnelly, C.A. (1998). Maternal transmission of BSE – interpretation of the data on the offspring of BSE-affected pedigree suckler cows. *Vet. Rec.*, **142**, 579–580.

Donnelly, C.A., Ferguson, N.M., Ghani, A.C., Wilesmith, J.W. and Anderson, R.M. (1997a). Analysis of dam–calf pairs of BSE cases: confirmation of a maternal risk enhancement. *Proc. Roy. Soc. London, Ser. B*, **264**, 1647–1656.

Donnelly, C.A., Ferguson, N.M., Ghani, A.C., Woolhouse, M.E.J., Watt, C.J. and Anderson, R.M. (1997b). The epidemiology of BSE in GB cattle herds: I. epidemiological processes, demography of cattle and approaches to control by culling. *Philos. Trans. Roy. Soc. London, Ser. B*, **352**, 781–801.

Donnelly, C.A., Ghani, A.C., Ferguson, N.M., Wilesmith, J.W. and Anderson, R.M. (1997c). Analysis of the bovine spongiform encephalopathy maternal cohort study: Evidence for direct maternal transmission. *Appl. Statist.*, **46**, 321–344.

Donnelly, C.A., Santos, R., Ramos, M., Galo, A. and Simas, J.P. (1999). BSE in Portugal: Anticipating the decline of an epidemic. *J. Epidemiology and Biostatistics, to appear*.

Durrett, R. and Levin, S. (1994a). The importance of being discrete (and spatial). *Theor. Pop. Biol.*, **46**, 363–394.

Durrett, R. and Levin, S. (1994b). Stochastic spatial models: a user's guide to ecological applications. *Philos. Trans. Roy. Soc. London, Ser. B*, **343**, 329–350.

Esslemont, R.J. (1992). Measuring dairy herd fertility. *Vet. Rec.*, **131**, 209–212.

Farquhar, C.F., Dornan, J., Somerville, R.A., Tunstall, A.M. and Hope, J. (1994). Effect of Sinc genotype, agent isolate and route of infection on the accumulation of protease-resistant PrP in non-central nervous system tissues during the development of murine scrapie. *J. Gen. Virol.*, **75**, 495–504.

Ferguson, N.M., Donnelly, C.A., Ghani, A.C. and Anderson, R.M. (1999a). Predicting the size of the epidemic of the new variant of Creutzfeldt-Jakob disease. *Brit. Food J.*, **101**, to appear.

Ferguson, N.M., Donnelly, C.A., Woolhouse, M.E.J. and Anderson, R.M. (1997a). The epidemiology of BSE in GB cattle herds: II. model construction and analysis of transmission dynamics. *Philos. Trans. Roy. Soc. London, Ser. B*, **352**, 803–838.

Ferguson, N.M., Donnelly, C.A., Woolhouse, M.E.J. and Anderson, R.M. (1997b). A genetic interpretation of heightened risk of BSE in offspring of affected dams. *Proc. Roy. Soc. London, Ser. B*, **264**, 1445–1455.

Ferguson, N.M., Donnelly, C.A., Woolhouse, M.E.J. and Anderson, R.M. (1999b). Estimation of the basic reproduction number of BSE: the intensity of transmission in British cattle. *Proc. Roy. Soc. London, Ser. B*, **266**, 23–32.

Ferguson, N.M., Ghani, A.C., Donnelly, C.A., Denny, G.O. and Anderson, R.M. (1998). BSE in Northern Ireland: epidemiological patterns past, present and future. *Proc. Roy. Soc. London, Ser. B*, **265**, 545–554.

Foster, J.D., Bruce, M., McConnell, I., Chree, A. and Fraser, H. (1996).

Detection of BSE infectivity in brain and spleen of experimentally infected sheep. *Vet. Rec.*, **138**, 546–548.

Foster, J.D. and Dickinson, A.G. (1988). Genetic control of scrapie in Cheviot and Suffolk sheep. *Vet. Rec.*, **123**, 159.

Foster, J.D., Hope, J., McConnell, I., Bruce, M. and Fraser, H. (1994). Transmission of bovine spongiform encephalopathy to sheep, goats and mice. *Ann. N.Y. Acad. Sci.*, **724**, 300–303.

Fraser, H., Bruce, M.E., Chree, A., McConnell, I. and Wells, G.A.H. (1992). Transmission of bovine spongiform encephalopathy and scrapie to mice. *J. Gen. Virol.*, **73**, 1891–1897.

Fraser, H., McConnell, I., Wells, G.A.H. and Dawson, M. (1988). Transmission of bovine spongiform encephalopathy to mice. *Vet. Rec.*, **123**, 472.

Fraser, H., Pearson, G.R., McConnell, I., Bruce, M.E., Wyatt, J.M. and Gruffydd-Jones, T.J. (1994). Transmission of feline spongiform encephalopathy to mice. *Vet. Rec.*, **134**, 449.

Gajdusek, D. and Zigas, V. (1957). Degenerative disease of the central nervous system in New Guinea: The endemic occurrence of kuru in the native population. *N. Engl. J. Med.*, **257**, 974–978.

Gajdusek, D.C. (1977). Unconventional viruses and the origin and disappearance of kuru. *Science*, **197**, 943–960.

Ghani, A.C., Ferguson, N.M., Donnelly, C.A., Hagenaars, T.J. and Anderson, R.M. (1998a). Epidemiological determinants of the pattern and magnitude of the vCJD epidemic in Great Britain. *Proc. Roy. Soc. London, Ser. B*, **265**, 2443–2452.

Ghani, A.C., Ferguson, N.M., Donnelly, C.A., Hagenaars, T.J. and Anderson, R.M. (1998b). Estimation of the number of people incubating variant CJD. *Lancet*, **352**, 1353–1354.

Gibson, G.J. (1997). Markov chain Monte Carlo methods for fitting spatiotemporal stochastic models in plant epidemiology. *Appl. Statist.*, **46**, 215–233.

Gibson, G.J. and Renshaw, E. (1998). Estimating parameters in stochastic compartmental models using Markov chain models. *IMA J. Math. Appl. Med. Biol.*, **15**, 19–40.

Goldmann, W., Hunter, N., Martin, T., Dawson, M. and Hope, J. (1991). Different forms of the bovine PrP gene have five or six copies of a short, G-C-rich element within the protein-coding exon. *J. Gen. Virol.*, **72**, 201–204.

Goldmann, W., Hunter, N., Smith, G., Foster, J. and Hope, J. (1994). PrP genotype and agent effects in scrapie: Change in allelic interaction with different isolates of agent in sheep, a natural host of scrapie. *J. Gen. Virol.*, **75**, 989–995.

Gordon, W.S., Brownlee, A. and Wilson, D.R. (1940). Studies of louping-ill, tick-borne fever and scrapie. In *Report of the Proceedings of the*

Third International Congress for Microbiology, pp. 362–363. Waverley, Baltimore.

Gore, S.M., Gilks, W.R. and Wilesmith, J.W. (1997). Bovine spongiform encephalopathy maternal cohort study — exploratory analysis. *Appl. Statist.*, **46**, 305–320.

Griffin, J.M., Collins, J.D., Nolan, J.P. and Weavers, E.D. (1997). Bovine spongiform encephalopathy in the Republic of Ireland: epidemiological observations 1989–1996. *Irish Vet. J.*, **50**, 593–600.

Hamer, W. (1906). Epidemic disease in England — the evidence of variability and of persistency of type. *Lancet*, **1**, 733–739.

Hartsough, G.R. and Burger, D. (1965). Encephalopathy of mink. I. Epizootiologic and clinical observations. *J. Infect. Dis.*, **115**, 387–392.

Hau, C.M. and Curnow, R.N. (1996). Separating the environmental and genetic factors that may be causes of bovine spongiform encephalopathy. *Philos. Trans. Roy. Soc. London, Ser. B*, **351**, 913–920.

Heesterbeek, J.A.P. and Dietz, K. (1996). The concept of R_0 in epidemic theory. *Statistica Neerlandica*, **50**, 89–110.

Heim, D. and Kihm, U. (1999). Bovine spongiform encephalopathy in Switzerland — the past and the present. *Rev. Sci. Tech. Off. Int. Epiz.*, **18**, 135–144.

Hill, A.F., Desbruslais, M., Joiner, S., Sidle, K.C.L., Gowland, I., Collinge, J., Doey, L.J. and Lantos, P. (1997a). The same prion strain causes vCJD and BSE. *Nature*, **389**, 448–450.

Hill, A.F., Zeidler, M., Ironside, J. and Collinge, J. (1997b). Diagnosis of a new variant Creutzfeldt-Jakob disease by tonsil biopsy. *Lancet*, **349**, 99–100.

Hilton, D.A., Fathers, E., Edwards, P., Ironside, J.W. and Zajick, J. (1998). Prion immunoreactivity in appendix before clinical onset of variant Creutzfeldt-Jakob disease. *Lancet*, **352**, 703–704.

Hoinville, L.J. (1996). A review of the epidemiology of scrapie in sheep. *Rev. Sci. Tech. Off. Int. Epiz.*, **15**, 827–852.

Hoinville, L.J., Wilesmith, J.W. and Richards, M.S. (1995). An investigation of risk factors for the cases of bovine spongiform encephalopathy born after the introduction of the feed ban. *Vet. Rec.*, **136**, 312–318.

Hope, J. (1994). The nature of the scrapie agent: the evolution of the virino. *Ann. N.Y. Acad. Sci.*, **724**, 282–289.

Hsiao, K. and Prusiner, S.B. (1990). Inherited human prion diseases. *Neurology*, **40**, 1820–1827.

Hunter, N., Dann, J.C., Bennett, A.D., Somerville, R.A., McConnell, I. and Hope, J. (1992). Are Sinc and the PrP gene congruent? evidence from PrP gene analysis in Sinc congenic mice. *J. Gen. Virol.*, **73**, 2751–2755.

Hunter, N., Foster, J.D., Dickinson, A.G. and Hope, J. (1989). Linkage of the gene for the scrapie-associated fibril protein (PrP) to the Sip gene in Cheviot sheep. *Vet. Rec.*, **124**, 364–366.

Hunter, N., Foster, J.D., Goldmann, W., Stear, M.J., Hope, J. and Bostock, C. (1996). Natural scrapie in a closed flock of Cheviot sheep occurs only in specific PrP genotypes. *Arch. Virol.*, **141**, 809–824.

Hunter, N., Goldmann, W., Smith, G., and Hope, J. (1994). Frequencies of PrP gene variants in healthy cattle and cattle with BSE in Scotland. *Veterinary Record*, **135**, 400–403.

Isham, V. (1989). Estimation of the incidence of HIV infection. *Philos. Trans. Roy. Soc. London, Ser. B*, **325**, 113–121.

Isham, V. (1995). Stochastic models of host−parasite interaction. *Ann. Appl. Prob.*, **5**, 720–740.

Jackson, G.S., Hosszu, L.L.P., Power, A., Hill, A.F., Kenney, J., Saibil, H., Craven, C.J., Waltho, J.P., Clarke, A.R. and Collinge, J. (1999). Reversible conversion of monomeric human prion protein between native and fibrilogenic conformations. *Science*, **283**, 1935–1937.

Keeling, M. and Grenfell, B. (1997). Disease extinction and community size: Modelling the persistence of measles. *Science*, **275**, 65–67.

Kermack, W.O. and McKendrick, A.G. (1927). Contributions to the mathematical theory of epidemics, Part I. *Proc. Roy. Soc., Ser. A*, **115**, 700–721. [Reprinted (1991) as *Bull. Math. Biol.*, **53**, 33-55.]

Kimberlin, R.H. (1982). Scrapie agent: prions or virinos? *Nature*, **297**, 107–108.

Kimberlin, R.H. (1990). Scrapie and possible relationships with viroids. *Semin. Virol.*, **1**, 153–162.

Kimberlin, R.H. and Walker, C.A. (1988). Pathogenesis of experimental scrapie. *Ciba Foundation Symp.*, **135**, 37–62.

Klitzman, R.L., Alpers, M.P. and Gajdusek, D.C. (1984). The natural incubation period of kuru and the episodes of transmission in three clusters of patients. *Neuroepidemiology*, **3**, 3–20.

Latarjet, R. (1979). Inactivation of the agents of scrapie, Creutzfeldt-Jacob disease and kuru by radiations. In Prusiner, S.B. and Hadlow, W.J., editors, *Slow Transmissible Diseases of the Nervous System*, Vol. 2, pp. 387–407. Academic Press, New York.

Marsh, R.F. and Bessen, R.A. (1994). Physiochemical and biological characterizations of distinct strains of the transmissible mink encephalopathy agent. *Philos. Trans. Roy. Soc. London, Ser. B*, **343**, 413–414.

Marsh, R.F., Bessen, R.A., Lehmann, S. and Hartsough, G.R. (1991). Epidemiological and experimental studies on a new incident of transmissible mink encephalopathy. *J. Gen. Virol.*, **72**, 589–594.

McKay, M.D., Beckman, R.J. and Conover, W.J. (1979). A comparison of three methods for selecting values of input variables in the analysis of output from a computer code. *Technometrics*, **21**, 239–245.

Medley, G.F., Anderson, R.M., Cox, D.R. and Billard, L. (1987). Incubation period of AIDS in patients infected via blood transfusion. *Nature*, **328**, 719–721.

Medley, G.F., Anderson, R.M., Cox, D.R. and Billard, L. (1988). Estimating the incubation period for AIDS patients. *Nature*, **333**, 504–505.

Medley, G.F.H. and Short, N.R.M. (1996). A model for the incubation period distribution of transmissible spongiform encephalopathies and predictions of the BSE epidemic in the United Kingdom. *Pre-print*.

Medori, R., Tritschler, H.J., LeBlanc, A., Villare, F., Manetto, V., Chen, H.Y., Xue, R., Leal, S., Montagna, P., Cortelli, P., Tinuper, P., Avoni, P., Mochi, M., Baruzzi, A., Hauw, J.J., Ott, J., Lugaresi, E., Autiliogambetti, L. and Gambetti, P. (1992). Fatal familial insomnia, a prion disease with a mutation at codon 178 of the prion protein gene. *N. Engl. J. Med.*, **326**, 444–449.

Middleton, D.J. and Barlow, R.M. (1993). Failure to transmit bovine spongiform encephalopathy to mice by feeding them with the extraneural tissues of affected cattle. *Vet. Rec.*, **132**, 545–547.

Moore, R.C., Hope, J., McBride, P.A., McConnell, I., Selfridge, J., Melton, D.W., and Manson, J.C. (1998). Mice with gene targetted prion protein alterations show that prnp, sinc and prni are congruent. *Nature Genetics*, **18**, 118–125.

Ministry of Agriculture, Fisheries and Food (1975–1990). *Agricultural Statistics – United Kingdom 1974–1989.* HMSO, London.

Ministry of Agriculture, Fisheries and Food (1992–1995). *The digest of agricultural census statistics – United Kingdom 1991–1994.* HMSO, London.

Ministry of Agriculture, Fisheries and Food (1996a). *Bovine Spongiform Encephalopathy in Great Britain: a Progress Report.* Ministry of Agriculture, Fisheries and Food.

Ministry of Agriculture, Fisheries and Food (1996b). *Programme to Eradicate BSE in the United Kingdom.* HMSO, London.

Morgan, K.L., Nicholas, K., Glover, M.J., and Hall, A.P. (1990). A questionnaire survey of the prevalence of scrapie in sheep in Britain. *Vet. Rec.*, **127**, 373–376.

National CJD Surveillance Unit and Dept of Infectious and Tropical Diseases, London School of Hygiene and Tropical Medicine (1998). *Creutzfeldt-Jakob Disease Surveillance in the UK, Seventh Annual Report.*

Neibergs, H.L., Ryan, A.M., Womack, J.E., Spooner, R.L. and Williams, J.L. (1994). Polymorphism analysis of the prion gene in BSE-affected and unaffected cattle. *Animal Genetics*, **25**, 313–317.

Oliver, R.M. and Smith, J.Q. (1990). *Influence diagrams, belief nets and decision analysis.* Wiley, New York.

Owen, F., Poulter, M., Collinge, J. and Crow, T.J. (1990). Codon 129 changes in the prion protein gene in Caucasians. *Am. J. Hum. Genet.*, **46**, 1215–1216.

Palsson, P.A. (1979). Rida (scrapie) in Iceland and its epidemiology. In Prusiner, S.B. and Hadlow, W.J., editors, *Slow Transmissible Diseases of the Nervous System*, Vol. 1, pp. 357–366. Academic Press, New York.

Pearson, G.R., Gruffydd-Jones, T.J., Wyatt, J.M., Hope, J., Chong, A., Scott, A.C., Dawson, M. and Wells, G.A.H. (1991). Feline spongiform encephalopathy. *Vet. Rec.*, **128**, 532.

Pearson, G.R., Wyatt, J.M., Gruffydd-Jones, T.J., Hope, J., Chong, A., Higgins, R.J., Scott, A.C. and Wells, G.A. (1992). Feline spongiform encephalopathy. *Vet. Rec.*, **131**, 307–310.

Press, W.H., Teukolsky, S.A., Vetterling, W.T. and Flannery, B.P. (1992). *Numerical Recipes in C.* Cambridge University Press, Cambridge.

Prusiner, S.B. (1982). Novel proteinaceous infectious particles cause scrapie. *Science*, **216**, 136–144.

Prusiner, S.B. (1991). Molecular biology of prion diseases. *Science*, **252**, 1515–1522.

Prusiner, S.B. (1996). Molecular biology and pathogenesis of prion diseases. *Trends in Biochemical Sci.*, **21**, 482–487.

Prusiner, S.B., Bolton, D.C., Groth, D.F., Bowman, K.A., Cochran, S.P., and McKinley, M.P. (1982). Further purification and characterization of scrapie prions. *Biochemistry*, **21**, 6942–6950.

Prusiner, S.B., Telling, G., Cohen, F.E. and DeArmond, S.J. (1996). Prion diseases of human and animals. *Semin. Virol.*, **7**, 159–173.

Purdey, M. (1992). Mad cows and warble flies: a link between BSE and organophosphates? *Ecologist*, **22**, 52–57.

Purdey, M. (1994). Are organophosphate pesticides involved in the causation of bovine spongiform encephalopathy (BSE)? hypothesis based upon a literature review and limited trials on BSE cattle. *J. Nutritional Med.*, **4**, 43–82.

Purdey, M. (1996a). The UK epidemic of BSE: Slow virus or chronic pesticide-initiated modification of the prion protein? Part 1: Mechanisms for a chemically induced pathogenesis/transmissibility. *Medical Hypotheses*, **46**, 429–443.

Purdey, M. (1996b). The UK epidemic of BSE: Slow virus or chronic pesticide-initiated modification of the prion protein? Part 2: An epidemiological perspective. *Medical Hypotheses*, **46**, 445–454.

Ramberg, J.S., Tadikamalia, P.R., Dudewicz, E.J. and Mykytka, E.F. (1979). A probability distribution and its uses in fitting data. *Technometrics*, **21**, 201–209.

Raymond, G.M., Hope, J., Kocisko, D.A., Priola, S.A., Raymond, L.D., Bossers, A., Ironside, J., Will, R.G., Chen, S.G., Petersen, R.B., Gambetti, P., Rubenstein, R., Smits, M.A., Lansbury, P.T. and Caughey, B. (1997). Molecular assessment of the potential transmissibilities of BSE and scrapie to humans. *Nature*, **388**, 285–288.

Ridley, R.M. and Baker, H.F. (1995). The myth of maternal transmission of spongiform encephalopathy. *Brit. Med. J.*, **311**, 1071–1075.

Ridley, R.M. and Baker, H.F. (1996a). No maternal transmission? *Nature*, **384**, 17.

Ridley, R.M. and Baker, H.F. (1996b). The paradox of prion disease. In Baker, H.F. and Ridley, R.M., editors, *Methods in Molecular Medicine: Prion Diseases*, pp. 1–13. Humana Press, Totowa, New Jersey.

Riesner, D., Kellings, K., Meyer, N., Mirenda, C. and Prusiner, S.B. (1992). Nucleic acids and scrapie prions. In Prusiner, S.B., Collinge, J., Powell, J. and Anderton, B., editors, *Prion diseases of humans and animals*, pp. 341–358. Ellis Horwood, London.

Rohwer, R.G. (1984a). Scrapie infectious agent is virus-like in size and susceptibility to inactivation. *Nature*, **308**, 658–662.

Rohwer, R.G. (1984b). Virus-like sensitivity of the scrapie agent to heat activation. *Science*, **223**, 600–602.

Schreuder, B. E.C., van Keulen, L.J.M., Vromans, M.E.W., Langeveld, J. P.M. and Smits, M.A. (1996). Preclinical test for prion diseases. *Nature*, **381**, 563.

Schreuder, B. E.C., van Keulen, L.J.M., Vromans, M.E.W., Langeveld, J. P.M. and Smits, M.A. (1998). Tonsillar biopsy and PrPSc detection in the preclinical diagnosis of scrapie. *Vet. Rec.*, **142**, 564–658.

The Scottish Office (1991–1995). *Economic report on Scottish agriculture 1990–1994*. HMSO, Edinburgh.

Simmons, M.M., Harris, P., Jeffrey, M., Meek, S.C., Blamire, I.W.H. and Wells, G.A.H. (1996). BSE in Great Britain: consistency of the neurohistopathological findings in two random annual samples of clinically suspect cases. *Vet. Rec.*, **138**, 175–177.

Southwood, T.R.E. (1978). *Ecological Methods*, 2nd edition Chapman & Hall, London.

Southwood, T.R.E., Epstein, M.A., Martin, W.B. and Walton, J. (1989). *Report of the Working Party on Bovine Spongiform Encephalopathy*. Department of Health/Ministry of Agriculture, Fisheries and Food.

Spongiform Encephalopathy Advisory Committee (1995). *Transmissible Spongiform Encephalopathies: a Summary of Present Knowledge and Research*. HMSO, London.

Spongiform Encephalopathy Advisory Committee (1997). *Advice to Ministers on dorsal root ganglia*.

Spraker, T.R., Miller, M.W., Williams, E.S., Getzy, D.M., Adrian, W.J.,

Schoonveld, G.G., Spowart, R.A., O'Rourke, K.I., Miller, J.M. and Merz, P.A. (1997). Spongiform encephalopathy in free-ranging mule deer (Odocoileus hemionus), white-tailed deer (Odocoileus virginianus) and Rocky Mountain elk (Cervus elaphus nelsoni) in north-central Colorado. *J. Wildlife Diseases*, **33**, 1–6.

Stamp, J.T. (1962). Scrapie: a transmissible disease of sheep. *Vet. Rec.*, **74**, 357–362.

Stamp, J.T. (1967). Scrapie and its wider implications. *Brit. Med. Bull.*, **23**, 133–137.

Stein, M. (1987). Large sample properties of simulations using Latin hypercube sampling. *Technometrics*, **29**, 143–151.

Stringer, S.M., Hunter, N. and Woolhouse, M.E.J. (1998). A mathematical model of the dynamics of scrapie within a sheep flock. *Math. Biosci.*, **153**, 79–98.

Tateishi, J., Hikita, K., Kitamoto, T. and Nagara, H. (1987). Experimental Creutzfeldt-Jakob disease: induction of amyloid plaques in rodents. In Prusiner, S.B. and McKinley, M.P., editors, *Prions: Novel Infectious Pathogens Causing Scrapie and Creutzfeldt-Jakob Disease*, pp. 415–426. Academic Press, New York.

Tateishi, J., Koga, M., Sato, Y. and Mori, R. (1980). Properties of the transmissible agent derived from chronic spongiform encephalopathy. *Ann. of Neurology*, **7**, 390–391.

Taylor, D.M. (1991). Spongiform encephalopathies. *Neuropathol. Appl. Neurobiol.*, **17**, 237.

Taylor, D.M. (1992). Inactiviation of unconventional agents of the transmissible degenerative encephalopathies. In Russell, A.D., Hugo, W.B. and Ayliffe, G.A.J., editors, *Principles and Practice of Disinfection, Preservation and Sterilization*, pp. 171–179. Blackwell, Oxford.

Taylor, D.M. (1994). Decontamination studies on the agents of bovine spongiform encephalopathy and scrapie. *Arch. Virol.*, **139**, 313–326.

Taylor, D.M. (1996). Exposure to and inactivation of, the unconventional agents that cause transmissible degenerative encephalopathies. In Baker, H.F. and Ridley, R.M., editors, *Methods in Molecular Medicine: Prion Diseases*, pp. 105–118. Humana Press, Totowa, New Jersey.

Thomas, P. and Newby, M. (1999). Estimating the size of the outbreak of new-variant CJD. *Brit. Food J.*, **101**, 44–58.

Weissmann, C. (1991a). The prion's progress. *Nature*, **349**, 569–571.

Weissmann, C. (1991b). A unified theory of prion propagation. *Nature*, **352**, 679–683.

Wells, G.A.H., Dawson, M., Hawkins, S.A.C., Green, R.B., Dexter, I., Francis, M.E., Simmons, M.M., Austin, A.R. and Horigan, M.W. (1994). Infectivity in the ileum of cattle challenged orally with bovine spongiform encephalopathy. *Vet. Rec.*, **135**, 40–41.

Wells, G.A.H., Hawkins, S.A.C., Green, R.B., Austin, A.R., Dexter, I., Spencer, Y.I., Chaplin, M.J., Stack, M.J. and Dawson, M. (1998). Preliminary observations on the pathogenesis of experimental bovine spongiform encephalopathy (BSE): an update. *Vet. Rec.*, **142**, 103–106.

Wells, G.A.H., Hawkins, S.A.C., Hadlow, W.J. and Spencer, Y.I. (1992). The discovery of bovine spongiform encephalopathy and observations on the vacuolar changes. In Prusiner, S.B., Collinge, J., Powell, J. and Anderton, B., editors, *Prion Diseases of Humans and Animals*, pp. 256–274. Ellis Horwood, Chichester.

Wells, G.A.H., Scott, A.C., Johnson, C.T., Gunning, R.F., Hancock, R.D., Jeffrey, M., Dawson, M. and Bradley, R. (1987). A novel progressive spongiform encephalopathy in cattle. *Vet. Rec.*, **121**, 419–420.

Wijeratne, W.V.S. and Curnow, R.N. (1990). A study of inheritance of susceptibility to bovine spongiform encephalopathy. *Vet. Rec.*, **126**, 5–8.

Wilesmith, J.W. and Ryan, J.B.M. (1997). Absence of BSE in the offspring of pedigree suckler cows affected by BSE in Great Britain. *Vet. Rec.*, **141**, 250–251.

Wilesmith, J.W., Ryan, J.B.M. and Atkinson, M.J. (1991). Bovine spongiform encephalopathy: epidemiological studies on the origin. *Vet. Rec.*, **128**, 199–203.

Wilesmith, J.W., Ryan, J.B.M. and Hueston, W.D. (1992a). Bovine spongiform encephalopathy: case-control studies of calf feeding practises and meat and bonemeal inclusion in proprietary concentrates. *Res. Vet. Sci.*, **52**, 325–331.

Wilesmith, J.W., Ryan, J.B.M., Hueston, W.D. and Hoinville, L.J. (1992b). Bovine spongiform encephalopathy: epidemiological features 1985–1990. *Vet. Rec.*, **130**, 90–94.

Wilesmith, J.W., Wells, G.A.H., Cranwell, M.P. and Ryan, J. B.M. (1988). Bovine spongiform encephalopathy: epidemiological studies. *Vet. Rec.*, **123**, 638–644.

Wilesmith, J.W., Wells, G.A.H., Hoinville, L.J. and Simmons, M.M. (1994). Suspected vertical transmission of BSE. (letter). *Vet. Rec.*, **134**, 198–199.

Wilesmith, J.W., Wells, G.A.H., Ryan, J. B.M., Gavier-Widen, D. and Simmons, M.M. (1997). A cohort study to examine maternally-associated risk factors for bovine spongiform encephalopathy. *Vet. Rec.*, **141**, 239–243.

Will, R.G., Cousens, S.N., Farrington, C.P., Smith, P.G., Knight, R.S.G., and Ironside, J.W. (1999). Deaths from variant Creutzfeldt-Jakob disease. *Lancet*, **353**, 979.

Will, R.G., Ironside, J.W., Zeidler, M., Cousens, S.N., Estibeiro, K., Alperovitch, A., Poser, S., Pocchiari, M., Hofman, A. and Smith, P.G.

6
(1996). A new variant of Creutzfeldt-Jakob disease in the UK. *Lancet*, **347**, 921–925.

Williams, E.S. and Young, S. (1980). Chronic wasting disease of captive mule deer: A spongiform encephalopathy. *J. Wildlife Diseases*, **16**, 89–98.

Williams, E.S. and Young, S. (1982). Spongiform encephalopathy of Rocky Mountain elk. *J. Wildlife Diseases*, **18**, 465–471.

Williams, E.S. and Young, S. (1992). Spongiform encephalopathies in Cervidae. *Rev. Sci. Tech. Off. Int. Epiz.*, **11**, 551–567.

Woolhouse, M.E.J. and Anderson, R.M. (1997). Understanding the epidemiology of BSE. *Trends in Microbiology*, **5**, 421–424.

Woolhouse, M.E.J., Matthews, L., Coen, P., Stringer, S.M., Foster, J.D., and Hunter, N. (1999). Population dynamics of scrapie in a sheep flock. *Philos. Trans. Roy. Soc. London, Ser. B*, **354**, 751–756.

Woolhouse, M.E.J., Stringer, S.M., Matthews, L., Hunter, N. and Anderson, R.M. (1998). Epidemiology and control of scrapie within a sheep flock. *Proc. Roy. Soc. London, Ser. B*, **265**, 1205–1210.

Zeidler, M., Stewart, G., Cousens, S.N., Estibeiro, K. and Will, R.G. (1997). Codon 129 genotype and new variant CJD. *Lancet*, **350**, 668.

Author index

Adrian, W.J., 7
Alper, T., 22
Alperovitch, A., 9, 14, 15
Alpers, M.P., 19
Anderson, R.M., 11, 16, 17, 19, 26, 28, 30, 31, 32, 34, 38, 39, 40, 41, 59, 60, 62, 63, 64, 69, 77, 81, 84, 86, 92, 98, 100, 103, 107, 111, 137, 155, 199
Atkinson, M.J., 10, 25
Austin, A.R., 10, 11, 16, 18, 19, 30, 62, 111, 137
Autiliogambetti, L., 9
Avoni, P., 9

Bacchetti, P., 37
Bailey, N.T.J., 42
Baker, H.F., 7, 10, 16
Barhen, J., 56
Barlow, R.M., 10
Baruzzi, A., 9
Bassett, H., 12
Bennett, A.D., 19
Berger, J.R., 7
Bessen, R.A., 7, 8
Birkett, C., 14
Blamire, I.W.H., 8
Bolker, B., 200
Bolton, D.C., 22
Bostock, C.J., 14, 21
Boughey, A.M., 9
Bovine Offal (Prohibition) Regulations, 9
Bovine Spongiform

Encephalopathy Order, 10, 25, 91
Bowman, K.A., 22
Boyd, A., 9
Bradley, R., 7, 10
Brookmeyer, R., 37
Brown, P., 7, 8, 9, 21, 22
Brownlee, A., 21
Bruce, M.E., 7, 8, 10, 14, 18, 19, 22, 198
BSE Inquiry, 10
Burger, D., 7, 22

Carolan, D.J.P., 12
Caughey, B., 22
Chandrakumar, M., 34
Chen, H.Y., 9
Chesebro, B., 22
Chong, A., 7
Chree, A., 8, 10, 14, 18, 19, 22, 198
Clarke, A.R., 22
Clarke, M.C., 22
Clements, R.A., 12
Cochran, S.P., 22
Coen, P., 199
Cohen, F.E., 19, 22
Collinge, J., 9, 14, 22
Collins, J.D., 12
Collins, S., 9
Comer, P.J., 197
Cortelli, P., 9
Cousens, S., 14
Cousens, S.N., 9, 14, 15, 34
Cox, D.R., 45, 154

Subject index